AF608844

ISNM
International Series of
Numerical Mathematics
Vol. 114

Edited by
K.-H. Hoffmann, München
H. D. Mittelmann, Tempe
J. Todd, Pasadena

Flow in Porous Media

Proceedings of the Oberwolfach Conference, June 21–27, 1992

Edited by

J. Douglas Jr.
U. Hornung

Springer Basel AG

Editors

Jim Douglas Jr.
Dept. of Mathematics
Purdue University
West Lafayette, IN 47907
USA

Ulrich Hornung
Fakultät für Informatik
Uni BwM
D-85577 Neubiberg
Germany

A CIP catalogue record for this book is available from the Library of Congress, Washington D.C., USA

Deutsche Bibliothek Cataloging-in-Publication Data
Flow in porous media : proceedings of the Oberwolfach conference, June 21–27, 1992 / ed. by J. Douglas ; U. Hornung. - Basel ; Boston ; Berlin : Birkhäuser, 1993
(International series of numerical mathematics ; Vol. 114)
ISBN 978-3-7643-2949-5 ISBN 978-3-0348-8564-5 (eBook)
DOI 10.1007/978-3-0348-8564-5
NE: Douglas, Jim [Hrsg.]; GT

Camera-ready copy prepared by the editors
Printed on acid-free paper produced from chlorine-free pulp
Cover design: Heinz Hiltbrunner, Basel

ISBN 978-3-7643-2949-5

9 8 7 6 5 4 3 2 1

TABLE OF CONTENTS

INTRODUCTION

Jim Douglas, Jr.*

These proceedings reflect some of the thoughts expressed at the Oberwolfach Conference on Porous Media held June 21–27, 1992, organized by Jim Douglas, Jr., Ulrich Hornung, and Cornelius J. van Duijn. Forty-five scientists attended the conference, and about thirty papers were presented. Fourteen manuscripts were submitted for the proceedings and are incorporated in this volume; they cover a number of aspects of flow and transport in porous media. Indeed, there are 223 individual references in the fourteen papers, but fewer than fifteen are cited in more than one paper.

The papers appear in alphabetical order (on the basis of the first author). A brief introduction to each paper is given below.

Allen and Curran consider a variety of questions related to the simulation of groundwater contamination. Accurate water velocities are essential for acceptable results, and the authors apply mixed finite elements to the pressure equation to obtain these velocities. Since fine grids are required to resolve heterogeneities, standard iterative procedures are too slow for practical simulation; the authors introduce a parallelizable, multigrid-based iterative scheme for the lowest order Raviart-Thomas mixed method. Contaminant transport is approximated through a finite element collocation procedure, and an alternating-direction, modified method of characteristics technique is employed to time-step the simulation. Computational experiments carried out on an nCube 2 computer.

Botz, Sternberg, and Greenspan analyze the fractal behavior of dispersion in miscible flow in porous media using both laboratory experiments and computer simulation based on a random walk model. They begin with laboratory experiments employing three linear, homogeneous, nonuniform porous media constructed in lucite columns with three sizes of spherical glass beads. The columns were joined end to end in various ways to create layered media. The effects of porosity, permeability, fluid velocity, viscosity, length, and layer order on dispersion were studied. Then, a random walk model was used to find heterogeneous dispersion coefficients by correlating the data through a multifractal analysis; the model predicts that dispersion is dependent on column length and order.

Bourgeat and Tapiéro apply homogenization to the Poisson equation on a domain including tiny holes periodically distributed with period diameter ε, except in a thin unperforated layer of thickness η. They study the behavior of the limit solution as ε and η tend to zero under different assumptions on the relation between ε and η. Proofs of the behavior are given using "two-scale convergence".

Cannon and DuChateau consider an inverse problem associated with unsaturated flow in porous media. An unknown coefficient in a quasilinear parabolic equation is

*Department of Mathematics, Purdue University, West Lafayette, IN 47907, USA

shown to be uniquely determined in a certain equivalence class by a combination of Dirichlet and Neumann data. They illustrate a coefficient reconstruction procedure for determining the coefficient, which they express directly in terms of overspecified data through an initial-boundary problem involving a trace-type functional partial differential equation. The algorithm associated with the reconstruction procedure was applied in some numerical experiments; inexact data was introduced to test the method.

Chadam studies the reaction-diffusion instability of reactive flows in porous media. Such flows alter the porosity and permeability of the medium and are modelled by means of a coupled system of nonlinear ordinary and partial differential equations. As the ratio of the equilibrium concentration of solute in the water to the density of the dissolving mineral tends to zero (this ratio is very small in realistic situations), the differential problem converges to a free boundary problem for the reaction interface. The stability of a planar reaction interface is discussed using local bifurcation analysis, and numerical results are presented to indicate the complexity of these interfaces as they evolve.

Douglas, Hensley, and Paes Leme generalize the medium block model introduced and considered earlier by Arbogast, Hornung, and the authors to allow a number of block types in the neighborhood of any point in the fracture sheet over the reservoir and to permit inhomogeneities in the physical parameters describing the system, after first outlining the derivation of the original medium block model for incompressible, immiscible displacement. A collection of numerical results are presented to study the effects of a variety of inhomogeneities on both a vertical cross-section problem and a five-spot problem.

Douglas, Paes Leme, Pereira, and Yeh consider a different numerical method for approximating the solution of the medium block model mentioned above. The object is to develop a computational algorithm in an easily parallelizable form. This is accomplished by replacing the finite difference approximation of the equations in the fractures by a mixed finite element procedure and then applying a version of a domain decomposition procedure originally introduced by Després for Helmholz problems; the subdomains consist of individual elements for the mixed method. The associated iterative procedure employed for each time step is explained, and then implementations for both SIMD and MIMD architectures are discussed. The effectiveness of the parallel codes is illustrated.

Islas and Lomen consider a solute transport problem related to the movement of pollutants and agricultural chemicals in soil water. They break the solute transport into two processes, convection and diffusion, and treat them independently. Convection is handled by the method of characteristics and diffusion by the method of singular perturbations. The underlying flux field is assumed known, so that the water content can be computed in advance of the calculation of solute concentration, allowing them to apply known analytical techniques to the two parts of the remaining problem for the solute transport.

Keyfitz shows that a number of systems modelling multiphase flow in a porous medium can be regarded as perturbations of conservation laws related to a class of flux functions which are hyperbolic except along a subspace of codimension one of state space. She describes change of type for these perturbations and discusses the existence of inaccessible regions in state space.

Kozlov studies the effective three-dimensional permeability in a heterogeneous medium having many length scales of heterogeneities. If the local permeability is assumed to be a random, statistically homogeneous field, then the homogenization theory of random partial differential equations with statistically homogeneous coefficients could be applied. Kozlov reformulates the assumption of statistical homogeneity to take multiple length scales into account in such a way as to lead to an effective permeability agreeing with recent experimental results of Noetinger and Jacquin, who predicted that the effective permeability should be given by the cube of the average of the third root of the local permeability.

Meyer considers the influence of surface tension on the stability of the interfaces in the Hele-Shaw and Muskat problems. Surface tension is seen through front-tracking calculations to mollify the cusping expected in the Hele-Shaw suction problem, which is unstable without surface tension. The Muskat problem is also unstable without surface tension when a more mobile fluid displaces a less mobile one. The addition of surface tension leads to damping of the higher frequency oscillations in the free surface. The front-tracking method employed in his calculations is carefully described.

Mikelic demonstrates the existence and uniqueness of a smooth solution for the Peaceman model for two-component, incompressible, miscible displacement in a porous medium under rather severe assumptions relating the viscosity function (or mobility ratio), the data functions, and the domain. The analysis is carried out in complete detail.

Showalter discusses several distributed microstructure models for flow in porous media, including two models for a totally fissured medium, others for a partially fissured medium allowing secondary flux between adjoining cells in addition to the predominate flow between the cells and the surrounding fissure system, and classical models realized as limiting cases of the microstructure models. For certain of these models, he indicates mathematical techniques for their analysis. There is an extensive and valuable bibliography cited for the models of this general type for porous media and several other classes of physical problems.

Su analyzes existence, uniqueness, and approximability of solutions for nonlinear, degenerate parabolic equations with evolutionary boundary conditions. Examples of physical problems described by systems of the type he treats include the simulation of the infiltration of rainfall through soil and a number of other unsaturated flow problems in porous media. Su considers a special case of the general problem and applies a parabolic regularization technique combined with some discrete schemes to carry him to existence and uniqueness of weak solutions. He also analyzes the discrete schemes as approximation methods.

International Series of Numerical Mathematics, Vol. 114, © 1993 Birkhäuser Verlag Basel

Parallelizable Methods for Modeling Flow and Transport in Heterogeneous Porous Media

Myron B. Allen* Mark C. Curran†

Abstract. Groundwater contaminant modeling presents several challenges to the mathematician. Among these are the need to compute accurate water velocities and difficulties arising from fine-scale heterogeneities and sharp concentration fronts. This paper presents parallelizable numerical methods that address these challenges.

For groundwater flow, mixed finite-element models yield velocities comparable in accuracy to computed heads. However, large variations in hydraulic conductivity can cause iterative matrix solvers to converge slowly. The fine grids needed to resolve heterogeneities aggravate the poor conditioning. A parallelizable, multigrid-based iterative scheme for the lowest-order mixed method largely overcomes both sources of poor behavior.

For contaminant transport, finite-element collocation yields high-order spatial accuracy. The timestepping scheme combines a modified method of characteristics, which reduces temporal errors when advection dominates, with an alternating-direction formulation, which is "embarrassingly parallel" and has a favorable operation count.

1 INTRODUCTION

The equations governing steady flow of water in a two-dimensional, rectangular porous medium Ω have the following forms:

$$\begin{aligned} u &= -K\,\nabla p \qquad \text{in } \Omega, \\ \nabla \cdot u &= f \qquad \text{in } \Omega. \end{aligned} \tag{1.1}$$

Here $u = (u^x, u^y), p$, and f represent the Darcy velocity, hydraulic head, and source term, respectively. In natural aquifers, the hydraulic conductivity $K(x, y)$ varies in space depending upon the lithology of the host rock. We assume that K is bounded above and that $\inf K(x, y) > 0$.

The spatial variability, or heterogeneity, in K causes difficulties in mathematical modeling. In particular, two sources of poor conditioning often affect the linear systems that approximate the governing equations. One source is the need to use fine spatial grids to resolve the variations in the medium and the resulting variations in p and u. The other is the variability in K itself, which affects the matrix entries of the linear system.

In this context, mixed finite-element methods have attracted much attention. These methods, together with appropriate choices of trial spaces, yield solutions for p and u

*Department of Mathematics, University of Wyoming, Laramie, WY 82071, USA (allen@corral.uwyo.edu)

†Applied and Numerical Mathematics Division, Sandia National Laboratories, Albuquerque, NM 87185, USA (mccurra@cs.sandia.gov)

that have the same order of accuracy as the grid mesh size $h \to 0$ ([5], [10]). Standard Galerkin and finite-difference formulations generally do not enjoy this property, since they require one to solve for p and then numerically differentiate to compute u. Since velocities determine the main features of the contaminant transport, mixed methods are, therefore, better suited to the coupled flow-and-transport problem.

Contaminant transport poses another set of difficulties. Here, the governing equation takes the form

$$\partial_t c + u \cdot \nabla c - \nabla \cdot (D_H \nabla c) = 0 \qquad \text{in } \Omega, \tag{1.2}$$

where $c(x,t)$ is the contaminant concentration and D_H represents the hydrodynamic dispersion tensor. This equation is formally parabolic.

In many applications, advection dominates, with the dissipative effects of hydrodynamic dispersion having only a small influence. In such regimes, the solution of (1.2) exhibits hyperbolic behavior, and sharp fronts in contaminant concentration tend to persist. Low-order numerical methods, such as upstream-weighted finite-differences, smear these fronts. Even high-order methods typically fail to capture the fronts accurately unless one uses either globally or locally fine spatial grids. In two or three space dimensions, the computational effort associated with such grids can be onerous, especially on serial-architecture machines.

Finite-element collocation on cubic trial spaces offers high-order spatial accuracy, but, like other techniques, it yields unwieldy matrix equations in the multidimensional problems arising in practice. An alternating-direction algorithm similar to that proposed by Celia [3] decomposes these unwieldy equations into parallelizable sets of smaller linear systems that can be solved with significantly fewer arithmetic operations. Moreover, the scheme is amenable to timestepping along approximate characteristic curves, a tactic that reduces the temporal truncation error; see Russell [11].

This paper examines these numerical methods. For the flow equations (1.1), we consider an iterative scheme for solving the lowest-order mixed finite-element approximations on rectangular grids. The overall structure of the scheme, analyzed in detail by Allen *et al.* [1], consists of an outer iteration, whose convergence rate is independent of h and of spatial variations in K, coupled with an inner iteration on an elliptic linear system. We use a highly parallelizable multigrid method to ensure that the inner iterations are rapid. For the transport equation (1.2), we examine an alternating-direction collocation (ADC) scheme that employs a modified method of characteristics and exhibits excellent parallelism (see Allen and Khosravani [2]).

2 THE MIXED-FINITE ELEMENT METHOD FOR THE FLOW EQUATIONS

Consider the equations (1.1), subject to the boundary condition $p = 0$ on $\partial\Omega$. To discretize this system via the lowest-order mixed method, we construct a rectangular grid Δ on Ω having vertical grid lines at $x = x_0, x_1, \ldots, x_m$ and horizontal grid lines at $y = y_0, y_1, \ldots, y_N$. The mesh size of Δ is $h := \max\{x_i - x_{i-1}, y_j - y_{j-1}\}$. With Δ we associate trial spaces Q_x, Q_y, and V for the x-velocity u^x, the y-velocity u^y, and the hydraulic head p, respectively. The space Q_x contains functions that are piecewise linear in x and piecewise constant in y; Q_y contains functions that are piecewise constant in x and piecewise linear in y, and V contains functions that are piecewise constant on Δ. Crucial to the error estimates associated with these spaces is the fact that, if $\mathbf{v} \in Q_x \times Q_y$, then $\nabla \cdot \mathbf{v} \in V$; see Raviart and Thomas [10].

Each of these trial spaces has a tensor-product basis containing products of the usual one-dimensional basis functions for piecewise constant and piecewise linear interpolation. We associate a nodal value $p_{i,j}$ of head with the centroid of each cell $[x_{i-1}, x_i] \times [y_{j-1}, y_j]$ formed by the grid Δ, a nodal value $u^x_{i,j}$ of x-velocity with the midpoint $(x_i, y_{j-1/2})$ of each vertical cell edge, and a nodal value $u^y_{i,j}$ with the midpoint $(x_{i-1/2}, y_j)$ of each horizontal cell edge.

Given these trial spaces, the mixed formulation for (1.1) is as follows: Find $u_h \in Q_x \times Q_y$ and $p_h \in V$ such that

$$\int_\Omega \frac{u_h \cdot \mathbf{v}}{K}\, dx\, dy - \int_\Omega p_h\, \nabla \cdot \mathbf{v}\, dx\, dy = 0, \quad \forall\, \mathbf{v} \in Q_x \times Q_y,$$

$$\int_\Omega (\nabla \cdot u_h - f) q\, dx\, dy = 0, \qquad \forall\, q \in V.$$

This finite-dimensional system yields approximations u_h and p_h whose global errors are both $O(h)$ in the norm $\|\cdot\|_{L^2(\Omega)}$; see [10].

Under lexicographic ordering of equations and unknowns, the equations (1.2) yield a linear system having the following block structure:

$$\begin{bmatrix} A & N \\ N^T & 0 \end{bmatrix} \begin{bmatrix} U \\ P \end{bmatrix} = \begin{bmatrix} 0 \\ F \end{bmatrix}. \tag{2.1}$$

The vector U contains nodal values of the velocities u^x and u^y, and P contains nodal heads. The matrix A is symmetric and positive definite and has the block structure

$$A = \begin{bmatrix} A^x & 0 \\ 0 & A^y \end{bmatrix}.$$

The blocks A^x and A^y are tridiagonal, their entries being integrals of the form

$$\int_\Omega K^{-1} \varphi_k \varphi_\ell\, dx\, dy,$$

where φ_k, φ_ℓ are functions belonging to the basis for $Q_x \times Q_y$. In practice, we approximate these integrals using a two-point Gauss composite rule in each coordinate direction.

The matrix N has the block structure

$$N = \begin{bmatrix} N^x \\ N^y \end{bmatrix},$$

where N^x and N^y. These blocks mimic the usual difference approximations to $\partial/\partial x$ and $\partial/\partial y$. The vector F contains integrals involving the source function f. For a detailed specification of the entries in this linear system, we refer readers to Allen *et al.* [1].

3 AN ITERATIVE SCHEME FOR THE MIXED METHOD

We solve the system (2.1) iteratively, using the following matrix splitting:

$$\begin{bmatrix} D & N \\ N^{\top} & 0 \end{bmatrix} \begin{bmatrix} U \\ P \end{bmatrix}^{(k+1)} = \begin{bmatrix} 0 \\ F \end{bmatrix} + \begin{bmatrix} D-A & 0 \\ 0 & 0 \end{bmatrix} \begin{bmatrix} U \\ P \end{bmatrix}^{(k)}. \tag{3.1}$$

Here, D is a diagonal matrix, the simplest effective structure for which is diag(A). This scheme has a convergence rate that is independent of mesh size and of variations in K. In fact, each iteration reduces the error by a factor not greater than 1/2 (see Allen *et al.* [1]).

Computationally, the scheme (3.1) requires the following steps:

(i) $G^{(k-1)} \leftarrow -F + N^{\top} D^{-1}(D-A)U^{(k-1)}$.

(ii) Solve $N^{\top} D^{-1} N P^{(k)} = G^{(k-1)}$.

(iii) $U^{(k)} \leftarrow D^{-1}(D-A)U^{(k-1)} - D^{-1}NP^{(k)}$.

Steps *(i)* and *(iii)* in this algorithm are cheap. Step *(ii)*, however, requires more work, since $N^{\top} D^{-1} N$ has the same pentadiagonal structure as the usual five-point finite-difference approximation to operators of the form $\nabla \cdot (K\nabla)$.

Instead of executing step *(ii)* exactly, we use a multigrid scheme to solve the pentadiagonal system approximately. Thus the matrix splitting serves as an "outer" iteration, while the multigrid cycles executed for step *(ii)* constitute an "inner" iteration. In particular, we perform several V-cycles to get an approximate value for $P^{(k)}$, then proceed to step *(iii)*. Each V-cycle involves two Gauss-Seidel iterations at each level in a nest $\Delta = \Delta_0 \supset \Delta_1 \supset \cdots \supset \Delta_L$ of successively coarser grids, the mesh size of Δ_k being $2^k h$. For the intergrid transfers, we use full weighting as a restriction operator and bilinear interpolation as a prolongation operator.

One attractive feature of the multigrid scheme is its amenability to parallel processing. Tuminaro and Womble [13], for example, discuss this advantage. By adopting a red-black ordering for the cells in each grid, we decompose each Gauss-Seidel relaxation sweep into two sets of calculations. In particular, we designate each cell $[x_{i-1}, x_i] \times [y_{j-1}, y_j]$ in a grid as *red* or *black*, depending on whether $i+j$ is even or odd. We update each of the red cells using old values in the black cells, then use the new red values to update the black cells. In any sweep, calculations for red cells are independent of each other. Updates for black cells are also mutually independent.

To implement the scheme on a distributed-memory machine, we arrange for each processor to manage a 32×32-cell rectangular region, or *patch*, of the original fine grid. The relaxation sweep on any patch requires some values of latest iterates from the nearest-neighbor patches. Therefore, before executing a relaxation sweep, a processor must trade information about a "boundary layer" of nodal values with the processor that manages the nearest-neighbor patch. Therefore, the parallel implementation requires communication between processors before each "red" sweep and before each "black" sweep. This communication prevents ideal parallel speedups.

4 COMPUTATIONAL PERFORMANCE OF THE MIXED-METHOD SCHEME

Allen *et al.* [1] discuss the performance of the serial scheme in the presence of the following heterogeneous conductivity fields $K(x, y)$ on $\Omega = (0, 1) \times (0, 1)$:

$$\begin{aligned} K_{\mathrm{I}}(x,y) &= 1; \\ K_{\mathrm{II}}(x,y) &= e^{-x-y}; \\ K_{\mathrm{III}}(x,y) &= \begin{cases} 1, & \text{if } x < y, \\ 0.1, & \text{if } x \geq y; \end{cases} \\ K_{\mathrm{IV}}(x,y) &= K_{\mathrm{II}}(x,y) \cdot K_{\mathrm{III}}(x,y); \\ K_{\mathrm{V}}(x,y) &= \begin{cases} 1, & \text{if } x < y, \\ 0.01, & \text{if } x \geq y. \end{cases} \end{aligned}$$

The experiments involve grids with $h = 2^{-\ell}$, where $\ell = 4, 5, 6, 7, 8$. Each iteration of the solution scheme includes two V-cycles of the multigrid algorithm, where the coarsest grid in each cycle has mesh 2^{-1}, and the finest has mesh $2^{-\ell}$. Table 1 displays the convergence rates of the outer iteration versus coefficient and mesh size. The results confirm the theoretical bound of 1/2 for the convergence rate.

Table 1: Convergence Rates for the Outer Iteration of the Flow-Equation Scheme Using Various Coefficients and Grids.

	Grid mesh h				
Coefficient	2^{-4}	2^{-5}	2^{-6}	2^{-7}	2^{-8}
K_{I}	0.4933	0.4988	0.4993	0.4995	0.4999
K_{II}	0.4966	0.4995	0.4988	0.4997	0.4999
K_{III}	0.4948	0.4982	0.4991	0.4998	0.4999
K_{IV}	0.4947	0.4980	0.4992	0.4998	0.4999
K_{V}	0.4939	0.4978	0.4989	0.4999	0.5000

To assess the scheme's parallelism, we examine its execution time on a 1024-processor nCube 2 having a hypercube architecture. To measure speedups, we examine execution times required on subcubes of the machine having dimension 0 (1 processor), 1 (2 processors), ..., 10 (1024 processors), running problems of proportionately larger size on larger subcubes. Each subcube is a set of processors linked by the shortest possible physical paths in the machine. Hence proper subcubes suffer essentially no disadvantage in the lengths of communication paths.

Table 2 shows timings for a sequence of runs involving a 32×32-cell grid on the one-processor subcube, a 64×32-cell grid on the two-processor subcube, a 64×64-cell grid on the four-processor subcube, and so forth, up to a 512×512 grid on a 512-processor cube. Since the ratio of problem size to number of processors remains constant in this sequence, an algorithm possessing ideal parallelism would require the same execution time for all runs. In practice, interprocessor communication and computational overhead disrupt this ideal relationship.

Table 2 also shows the times associated with problem setup (initialization and matrix assembly) and interprocessor communication. Each run represents 20 outer iterations of the scheme (2.1), each iteration of which requires five V-cycles in step *(ii)*. In practice, the outer iterations typically converge to within machine precision tolerances in fewer than 10 iterations, so practical runtimes are smaller, and setup time has a larger effect on speedup. Still, these timings suggest that the algorithm possesses excellent parallelism in addition to its good performance in the presence of heterogeneities and fine grids.

Table 2: Run Times (seconds) for Scaled Groundwater Flow Problems on the nCUBE 2.

Number of processors	Setup time	Communication time	Total time
1	0.178	0.205	161.516
2	0.184	1.784	159.274
4	0.201	3.690	157.260
8	0.214	4.639	158.484
16	0.251	5.775	159.967
32	0.319	5.832	160.231
64	0.530	5.908	160.620
128	0.782	5.950	160.985
256	1.396	5.979	161.673
516	2.691	6.030	163.005

5 COLLOCATION FOR THE TRANSPORT EQUATION

We turn now to (1.2), which governs contaminant transport. Of special interest are flow regimes in which advection is dominant, in the sense that, if L is the diameter of the spatial domain, then the Peclet number $\|u\|_\infty L/D_H$ is much larger than unity. For such problems, it is useful to rewrite (1.2) in terms of the material derivative $D_t := \partial_t + u \cdot \nabla$ of the fluid-solute mixture. We get

$$D_t c \;-\; \nabla \cdot (D_H \nabla c) \;=\; 0\,. \tag{5.1}$$

Consider the following initial-boundary-value problem:

$$\begin{aligned} D_t c \;-\; \nabla \cdot (D_H \nabla c) &= 0, \quad (x,t) \in \Omega \times (0,\infty), \\ c(x,0) &= c_I(x), \quad x \in \Omega, \\ c(x,t) &= 0, \quad (x,t) \in \partial\Omega \times (0,\infty). \end{aligned}$$

This problem models the movement of an initial contaminant plume $c_I(x)$, so long as the plume does not approach $\partial\Omega$.

To discretize this problem in space, we use finite-element collocation on piecewise Hermite bicubics, a standard method summarized, for example, in Curran and Allen [4]. Let Δ be a rectangular grid partitioning Ω into rectangular elements bounded by adjacent grid lines $x = x_i$ and $y = y_j$. As before, h stands for the mesh size of this grid. Denote by M the trial space of all Hermite piecewise bicubics that vanish on $\partial\Omega$. The trial function $c_h \in M$ has the form

$$c_h = \sum_{i,j} \left(c_{ij} H_{00ij} + c_{ij}^{(x)} H_{10ij} + c_{ij}^{(y)} H_{01ij} + c_{ij}^{(x,y)} H_{11ij} \right),$$

where the functions $H_{pqij}(x, y)$ form a nodal basis for M; see [9].

To determine the nodal unknowns in this expansion, we substitute c_h into the left side of Equation (5.1) and force the residual to vanish at a set of collocation points $\bar{x}_m$. For optimal-order accuracy, we choose these points to be the 2×2 Gauss quadrature abscissae in each element Ω_i. This procedure yields a system of ordinary differential equations in time:

$$D_t c_h(\bar{x}_m, t) \; - \; \nabla \cdot [D_H \nabla c_h(\bar{x}_m, t)] \; = \; 0. \tag{5.2}$$

These equations determine the evolution of the unknown coefficients of c_h. We project the initial function c_I onto M via interpolation to get an initial function $c_h(\bar{x}_m, 0)$.

We discretize (2.1) temporally in two steps. First, following Russell [11], we approximate $D_t c_h$ using the modified method of characteristics (MMOC). This procedure leads to a difference expression of the form

$$D_t c_h(\bar{x}_m) \; \simeq \; k^{-1} \left[c_h^{n+1}(\bar{x}_m) - c_h^n(x_m^*) \right] ,$$

where $c_h^n(x)$ denotes an approximate value of $c_h(x, nk)$ and k is the time step. The point x_m^* is a *backtrack* point, which we compute according to the method of characteristics for the purely advective version of (1.2). Theoretically, if $(\mathbf{s}(t), t)$ is a parametrization of the characteristic curve $dx/dt = u$ passing through $\bar{x}_m$, then

$$x_m^* \; = \; \bar{x} \; + \; \int_{t_{n+1}}^{t_n} u(\mathbf{s}(t), t) \, dt.$$

In practice we compute x_m^* approximately by solving $dx_m/dt = u$, subject to the "final" condition $x(t_n) = \bar{x}_m$, using an Euler scheme.

The second step in discretizing (5.2) is to use alternating-direction collocation. We perturb the discrete operator equations to obtain the following factoring along the x- and y-coordinate directions:

$$(1 + k\mathcal{L}_x)(1 + k\mathcal{L}_y) c_h^{n+1}(\bar{x}_m) \; = \; c_h^n(x_m^*) + \mathcal{O}(k^2) \, . \tag{5.3}$$

Here, $\mathcal{L}_x = -\partial_x(D_H \partial_x)$ and $\mathcal{L}_y = -\partial_y(D_H \partial_y)$. By properly numbering the collocation equations and unknowns, one can reduce the equations (5.3) to an algebraic system that involves highly parallel sets of matrix equations, each of which has an inexpensive, one-dimensional structure.

6 COMPUTATIONAL ASPECTS OF ADC

Curran and Allen [4] discuss efficient algorithms for solving the ADC equations on parallel-architecture computers. The computational problem is "embarrassingly parallel," in the sense that it naturally decomposes into linear systems, having one-dimensional zero structure, that one can obviously solve concurrently. Speedup curves of slope greater than 0.8 are attainable on an Alliant FX/8 eight-processor machine.

Aside from parallelism, two features of the ADC-MMOC approach make it an attractive one. First, the method inherits high-order spatial accuracy from the standard collocation approach. Percell and Wheeler [8] show that standard collocation on piecewise Hermite cubics has $\mathcal{O}(h^4)$ spatial accuracy for elliptic spatial operators. ADC attains this accuracy with "one-dimensional" matrices having bandwidth five.

Second, the use of MMOC reduces both the temporal truncation error and the number of degrees of freedom needed to resolve sharp fronts. Russell [11] discusses these advantages. A related observation that MMOC essentially removes the advective term from the spatial operator, leaving only the diffusive operator to be discretized via collocation. This fact is appealing on numerical grounds, since we expect collocation on Hermite cubics to yield $\mathcal{O}(h^4)$ accuracy for (1.2) in the parabolic case, when $D_H \neq 0$, but only $\mathcal{O}(h^3)$ accuracy in the hyperbolic case when $D_H = 0$ (see [6]). With MMOC, the collocation procedure discretizes the part $-\nabla \cdot (D\nabla)$ of the spatial operator for which it is best suited, even when the other term $u \cdot \nabla$ is physically dominant.

The ADC-MMOC scheme does not strictly conserve mass in the global sense

$$E_M(t_n) := \int_\Omega \left(c_h^n - c_h^0\right) dv + \sum_{\nu=0}^{n} k \oint_{\partial\Omega} (uc_h^\nu - D\nabla c_h^\nu) \cdot \mathbf{n}\, ds = 0. \tag{6.1}$$

This effect is common in Eulerian-Lagrangian methods [11], [7]. Numerical experiments indicate, however, that the mass balance errors are typically not excessive. In a rotating plume problem on $\Omega = (-1,1) \times (-1,1)$ and $T = 1$, with $h = 0.02$, the mass balance error varies with the time step k. Table 3 shows values of the *relative mass balance error*,

$$R_M := \frac{|E_M(1)|}{\int_\Omega c_h^0\, dv}, \tag{6.2}$$

for four choices of k. Since accurate backtracking is necessary to obtain reasonable mass balance, the table also shows the number N_E of Euler steps used to compute the backtrack points x_m^* in each case.

7 DISCUSSION

A variety of extensions are needed to make these numerical methods fully useful in modeling porous-media flows. The most obvious needs are to extend the scheme for the flow equation to time-dependent, three-dimensional settings and to extend the ADC-MMOC scheme for the contaminant transport equation to three dimensions. These extensions involve modifications that, while conceptually straightforward, require nontrivial changes to the codes and will result in more computationally intensive algorithms. The principles that allow parallelizations should remain intact, however,

Table 3: Relative Mass Balance Errors R_M in the ADC-MMOC Scheme for a Rotating-Plume Problem on $\Omega = (-1,1) \times (-1,1)$, with $h = 0.02$ and $T = 1$. N_E is the Number of Euler Steps Used in the Backtracking.

Time step k	N_E	R_M
0.02	10	0.018
0.01	5	0.007
0.005	2	0.027
0.0025	2	0.015

so the approaches described here should be even more attractive in higher-dimensional applications.

More interesting is the need to extend the methods to problems involving tensor conductivities and tensor hydrodynamic dispersion. It is in the context of tensor conductivities that the two-level iterative scheme for the mixed-method equation has the greatest potential for practical use. Shen [12], through delicate analysis, shows that one can lump the matrix A in the mixed-method system and preserve global accuracy in the scalar case. Thus one can eliminate the need for the outer iterations used here. However, the analysis does not appear to extend to the case when the conductivity K is a tensor. In this case, the inner-outer iteration scheme still offers reasonable prospects for effective parallelism.

Incorporating tensor hydrodynamic dispersion into the ADC-MMOC formalism most likely will require an iterative formulation, in which one lags off-diagonal entries of D_H by an iteration. The use of iterations in this setting opens the way for simultaneous iterative reduction of the truncation error introduced in the operator splitting used to effect the alternating-direction strategy. The parallelism inherent in the ADC-MMOC approach makes iterations affordable.

The overall approach of combining alternating-direction techniques with the MMOC is by no means restricted to finite-element collocation. Krishnamachari *et al.* [7] discuss a related approach for a Galerkin scheme using piecewise bilinear trial functions, and one can easily imagine analogous schemes involving finite differences.

Acknowledgments. The Wyoming Water Research Center supported this work in part through a grant-in-aid. The U.S. National Science Foundation provided support through grant number EHR-910-8774. This work received support from the Applied Mathematical Sciences Program, U.S. Department of Energy Office of Energy Research. The work was performed in part at Sandia National Laboratories for the U.S. DOE under contract number DE-AC04-76DP00789. The first author expresses gratitude to the College of Engineering and Mathematics at the University of Vermont, which provided valuable computer time and expertise during a sabbatical leave.

REFERENCES

[1] Allen M. B., Ewing R. E., Lu P. Well conditioned iterative schemes for mixed finite-

element models of porous-media flow. *SIAM Jour. Sci. Stat. Comp.*, 13:794–814, 1992.

[2] Allen M. B., Khosravani A. Solute transport via alternating-direction collocation using the modified method of characteristics. *Advances in Water Resources*, 15:125–132, 1992.

[3] Celia M. A. *Collocation on deformed finite elements and alternating direction collocation methods.* PhD thesis, Princeton University, Princeton, 1983.

[4] Curran M. C., Allen M. B. Parallel computing for solute transport models via alternating-direction collocation. *Adv. Water Resour.*, 13(2):70–75, 1990.

[5] Douglas J., Ewing R. E., Wheeler M. F. The approximation of the pressure by a mixed method in the simulation of miscible displacement. *R.A.I.R.O. Analyse Numerique*, 17:17–33, 1983.

[6] Dupont T. Galerkin methods for first-order hyperbolics: An example. *SIAM J. Numer. Anal.*, 10:890–899, 1973.

[7] Krishnamachari S. V., Hayes L. J., Russell T. F. A finite element alternating-direction method combined with a modified method of characteristics for convection-diffusion problems. *SIAM J. Numer. Anal.*, 26(6):1462–1473, 1989.

[8] Percell P., Wheeler M. F. A C^1 finite element collocation method for elliptic equations. *SIAM J. Numer. Anal.*, 17(5):605–622, 1980.

[9] Prenter P. M. *Splines and Variational Methods.* Wiley, New York, 1975.

[10] Raviart P. A., Thomas J. M. A mixed finite element method for second order elliptic problems. In *Mathematical Aspects of the Finite Element Method*, volume 606 of *Lecture Notes in Mathematics*, pages 292–315. Springer-Verlag, Berlin and New York, 1977. I. Galligani and E. Magenes, eds.

[11] Russell T. F. *An incompletely iterated characteristic finite element method for a miscible displacement problem.* PhD thesis, University of Chicago, Chicago, 1980.

[12] Shen J. *Mixed finite element methods: analysis and computational aspects.* PhD thesis, University of Wyoming, Laramie, 1992.

[13] Tuminaro R. S., Womble D. E. Analysis of the multigrid FMV cycle on large-scale parallel machines. *SIAM Jour. Sci. Stat. Comp.* To appear.

International Series of Numerical Mathematics, Vol. 114, © 1993 Birkhäuser Verlag Basel

A MULTIFRACTAL ANALYSIS OF DISPERSION DURING MISCIBLE FLOW IN POROUS MEDIA

Michael M. Botz* Steven P. K. Sternberg* Robert A. Greenkorn*

Abstract. Dispersion experiments were performed using three linear, homogeneous, non-uniform porous media constructed in lucite columns with three sizes of spherical glass beads as the porous media. The columns were joined end to end to create a series of layered heterogeneous porous media. Each column and all combinations of columns were studied to determine the effects that porosity, permeability, velocity, viscosity, length, and column order have upon dispersion.

A computer simulation based on a random walk model was used to predict the heterogeneous dispersion coefficients using data from the homogeneous columns. The model correlates the data using a multifractal analysis and predicts that dispersion is dependent upon length and column order.

1 INTRODUCTION

Accurate prediction of dispersion in natural and man-made porous media is important in many fields. Current prediction models are suitable for many man-made and homogeneous media, but in heterogeneous and large-scale natural systems such as aquifers and oil fields, current predictive methods are inadequate.

The conventional advection-dispersion equation, which in one dimension is given by

$$\frac{\partial C}{\partial t} = \mathcal{D}\frac{\partial^2 C}{\partial x^2} - v\frac{\partial C}{\partial x}, \tag{1.1}$$

has been solved (see, e.g., [13]) for a step function input of tracer. Equation (1.1) provides satisfactory concentration profiles for dispersion in uniform homogeneous porous media, but is inadequate to model dispersion in heterogeneous media [14], [16], [1]; [17], [10], [11].

Deterministic dispersion models are most often based on some form of averaging the microscopic equations over a hierarchy of discrete scales [2]. The averaging techniques rely upon the periodicity of the fine structure; however, this assumption is rarely met. Dagan [5] concludes the field scale will not, in general, be representable as a Fickian process. Fickian transport results when there is sufficient mixing between random velocities and when velocity correlations are relatively short-ranged compared with the solute travel distance. Except for very special cases, true transport processes do not satisfy the advection-dispersion equation [3].

Dispersivity appears to increase with length [6], [12], but it is asymptotically constant when the flow path is very large compared to the integral scale of permeability

*Department of Chemical Engineering, Purdue University, West Lafayette, IN

distribution [4], [9]. The increase in dispersivity with length is called the scale effect; scale-dependent dispersivity has been treated using concepts from fractal geometry [12], [17] and can lead to large differences in measured values of dispersivity between laboratory and field scales.

The object of this work is to study dispersion during miscible flow in a heterogeneous porous medium composed of a series of non-uniform but homogeneous layers with differing permeabilities. The terms non-uniform and homogeneous are used throughout this paper in the manner described by Greenkorn and Kessler [8].

2 DISPERSION

Dispersion during miscible flow in porous media is defined by

$$\mathcal{D} = \frac{\sigma^2}{2t}, \tag{2.1}$$

where $\mathcal{D}$ is the dispersion coefficient $[L^2/T]$, σ^2 is the variance of the concentration in the dispersing front $[L^2]$ and t is the time of variation $[T]$. Dispersion is commonly determined using concentration versus time data, in which case dispersion is determined using

$$\mathcal{D} = \frac{\sigma^2 L v}{2t_{0.50}^2}, \tag{2.2}$$

where L is the path length $[L]$, v is the advective velocity $[L/T]$, $t_{0.50}$ is the breakthrough time corresponding to a relative concentration of 0.50, and σ^2 is the variance in arrival time given by

$$\sigma = \frac{t_{0.8413} - t_{0.1587}}{2}.$$

3 EXPERIMENTAL PROCEDURE AND RESULTS

3.1 Description of dispersion in the laboratory

For a complete description of the experimental apparatus and procedure see Sternberg [15]. The porous media were constructed using 30.5 cm long, 3.2 cm inside diameter, lucite columns packed with spherical glass beads. The lucite columns are made such that they can be connected with one another in all combinations. A 0.20M potassium chloride solution was used as the tracer fluid, and silver/silver chloride electrodes were used to determine transient tracer concentrations within the columns. Fluid velocities ranged from 0.011 cm/sec to 0.027 cm/sec. The physical properties of the three media investigated are shown in Table 1.

3.2 Laboratory results

Concentration versus time data were collected during the dispersion experiments for the three columns and for several combinations and permutations of the columns, each at three fluid velocities. Dispersion coefficients were determined using (2.2).

Because dispersion coefficients are a function of velocity, dispersivity is often used to describe the mixing in porous media; dispersivity is independent of the advective

Table 1: Description of the Porous Media

Column Number	Avg. Bead Size	Bead Range	Porosity	Permeability
	mm	mm	%	darcy
1	0.60	0.84 – 0.42	33.6	233
2	0.25	0.42 – 0.21	32.5	51
3	0.15	0.21 – 0.11	33.7	14

velocity. Dispersivity is the slope of a best-fit line on a plot of dispersion versus velocity or some appropriate function of velocity such as a power law relationship, as used in this study. Dispersion is then related to the fluid velocity by

$$\mathcal{D} = \alpha v^m, \tag{3.1}$$

where α is the dispersivity $[L]$, and $m = 1.2$; see [7]. The results of the dispersivity experiments for the non-uniform homgeneous columns and the combinations of columns forming the heterogeneous media are given in Table 2. The column numbers in the table are deciphered as follows: Each digit in the column number represents one of the columns. A series of digits represent a model consisting of more than one column, with the order of the digits corresponding to the order in which the tracer fluid encounters the columns. Thus, Model 13 consists of two columns, 1 and 3, arranged such that the tracer fluid flows through Column 1 followed by Column 3.

As expected, the dispersivity increases with the glass bead diameter; it was also found to be very sensitive to changes in temperature. Dispersion coefficients were found to be a function of permeability, advective velocity, viscosity, length, and column order. Concentration profiles generated using the advection-dispersion equation did not agree with the experimentally observed profiles.

4 RANDOM WALK MODEL

4.1 Description of the model

Random walk streamlines were created and used to approximate the motion of tracer particles as they flow through a porous medium. The particles were randomly moved a given length along an axis, and the dispersion coefficient was determined from the spread of the particles.

A large number of particles $(200 - 500)$ were positioned at the $x - y$ origin at time zero. Then, each particle was subjected to random steps of the form

$$x_{i+1} = x_i + v_x + \delta * R_x, \qquad y_{i+1} = y_i + \delta * R_y, \tag{4.1}$$

where v_x was the specified average velocity in the x-direction, R_x and R_y were random numbers uniformly distributed between -1 and $+1$, and δ was the maximum allowable jump distance. The parameter δ was taken to be the same for the x- and y-directions. Figure 1 illustrates the movement of a particle in a single step. Perturbing the particles randomly within the square is the source of dispersion. Each particle is repeatedly

Table 2: Experimental Dispersivities

Column Number	Dispersivity mm
1	0.611±0.002
2	0.244±0.001
3	0.183±0.002
12	0.574±0.001
21	0.587±0.002
23	0.238±0.001
32	0.174±0.001
13	0.557±0.001
31	0.447±0.002
123	0.247±0.001
321	0.271±0.001
132	0.449±0.003
231	0.296±0.002

subjected to (4.1) until the value of x_{i+1} is at least equal to the specified length of the system. The total actual length of the particle path and the net distance traveled in the x-direction are used to determine the streamline's fractal dimension.

The fractal dimension of the particle traces are determined using results from Wheatcraft and Tyler [17],

$$L_f = L_s^d \varepsilon^{1-d}, \tag{4.2}$$

where L_f is the actual path length, L_s is the net distance moved in the x-direction, ε is the fractal cutoff, and d is the fractal dimension. Since

$$\ln L_f = d \ln L_s + (1-d) \ln \varepsilon, \tag{4.3}$$

d can be found by determining the slope of a plot of $\ln L_f$ versus $\ln L_s$; it has already been noted that the slope is independent of the value chosen for ε. (See [17] for a discussion concerning the fractal cutoff.)

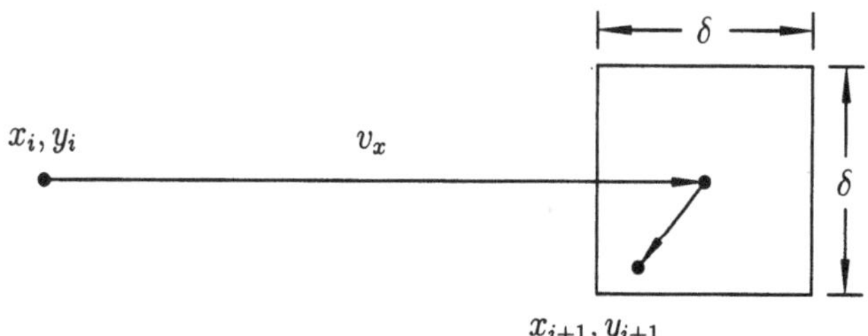

Figure 1: Illustration of Particle Movement

Table 3: Random Walk Dispersivities

Column Number	Dispersivity (experimental) mm	Dispersivity (random model) mm
1	0.611	0.64
2	0.244	0.26
3	0.183	0.16
12	0.573	0.47
21	0.587	0.44
23	0.238	0.19
32	0.174	0.19
13	0.557	0.34
31	0.447	0.48
123	0.247	0.35
321	0.271	0.34
132	0.449	0.30
231	0.296	0.28

The random walk model is not a deterministic fractal model; that is, the fractal dimension is not specified at the beginning of a walk but is determined after the walk is completed.

4.2 Random walk dispersion results

Using the results of the laboratory dispersion experiments for the individual non-uniform homogeneous columns, dispersion coefficients for the heterogeneous columns were computed. This was done by adjusting the parameter δ by trial and error in (4.1) so that the dispersion of the random walk model closely matched the dispersion in the individual columns that was observed in the laboratory. Nine values of δ were needed to match each bead size and velocity. For each bead size a second order polynomial was used to correlate δ with fluid velocity, thus allowing input of any velocities within the experimental range to the model. In this manner, the random walk model was fitted to the homogeneous dispersion data for all three columns and used to predict the heterogeneous dispersion coefficients.

As remarked above, fractal dimensions are calculated by determining the slope of a plot of $\ln L_f$ versus $\ln L_s$. Heterogeneous dispersion coefficients were then computed by piecing together several random walks corresponding to the order of columns of interest. The addition of the individual columns to form a series of homogeneous layers results in a multifractal random walk simulating a heterogeneous porous medium. An overall fractal dimension from each of the random walks was computed using (4.3). By performing the experiments at three different velocities, dispersivities were calculated using (3.1). The random model dispersivity results are compared to the experimental dispersivity results in Table 3. As shown, the individual column data are nearly the same for both sets of results because the data from the laboratory and the random

Table 4: Fractal Dimensions of Particle Traces

Column Number	Velocity cm/sec	Fractal Dimension
1	0.0112	1.3046
	0.0160	1.2875
	0.0256	1.2305
2	0.0115	1.2169
	0.0165	1.1781
	0.0263	1.1362
3	0.0112	1.1558
	0.0160	1.1413
	0.0255	1.0942
12	0.0115	1.2173
	0.0164	1.1965
	0.0262	1.1556
21	0.0115	1.2114
	0.0164	1.1896
	0.0263	1.1480
123	0.0114	1.1731
	0.0163	1.1556
	0.0261	1.1192
321	0.0114	1.1682
	0.0163	1.1502
	0.0261	1.1135

models were forced to match in these cases. It is also seen that the columns do not exhibit reciprocity; i.e., the dispersion coefficients are dependent upon the order in which the fluid encounters the columns.

The fractal dimensions determined from the particle streamlines for the three individual columns and for some of the combinations of the columns are shown in Table 4. The fractal dimensions were found to increase with particle diameter and decrease with advective velocity. In the case of the heterogeneous columns, one fractal dimension was computed for the entire random walk, although the random walk was actually multifractal. By multifractal, it is meant that the entire walk is fractal, but is composed of a subset of fractal walks, each with a unique fractal dimension. In the case of three columns, the multifractal walk is composed of a subset of three fractal walks. Random walk streamlines for the sets of three columns are shown in Figure 2. The random walk streamlines shown in Figure 2 were made using the same sequence of random numbers so as to allow the comparison of the walks. Each plot corresponds to a different sequence of three columns, each with a length of 30.5 cm. By comparing respective

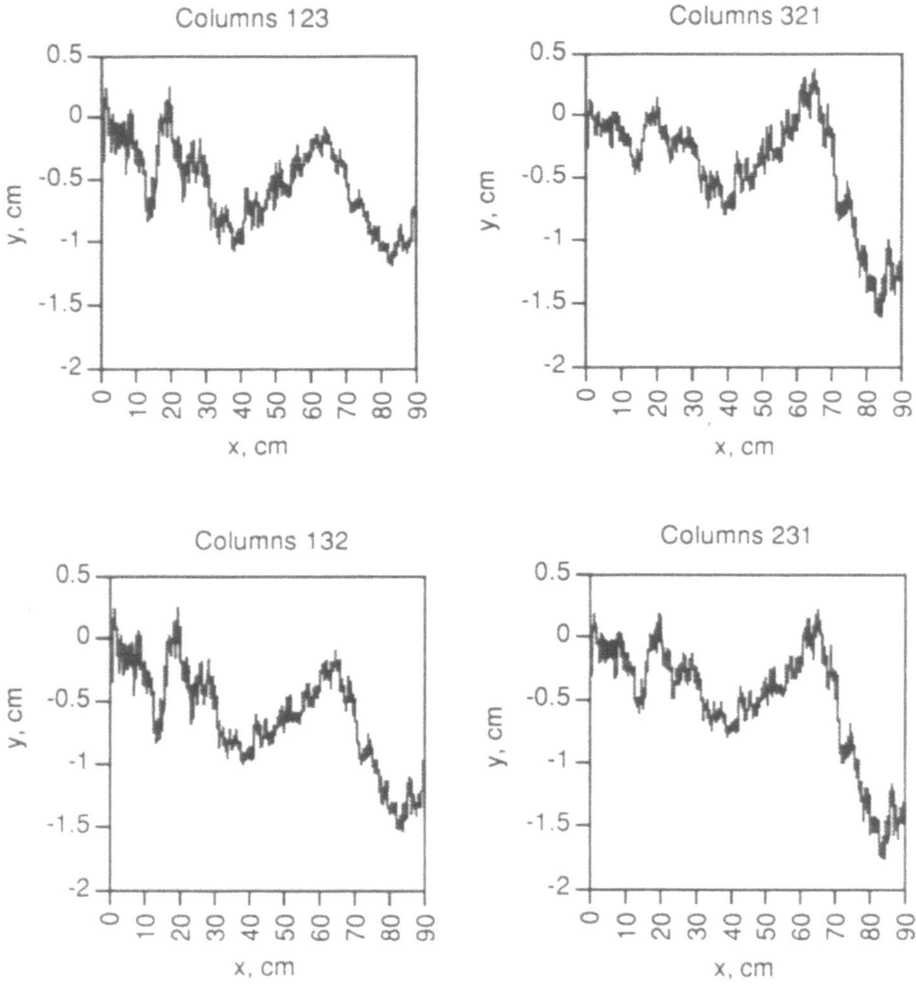

Figure 2: Random Walk Streamlines for the Sets of Three Columns at the Same Fluid Velocity

one-thirds of the traces, differences between the walks can be observed.

5 CONCLUSIONS

The results from the laboratory experiments of Sternberg [15] showed that dispersion is a function of permeability, advective velocity, viscosity, length, and column order. The dependence of dispersion upon column order is an unexpected result.

The use of a multifractal random walk model which is composed of a union of a discrete number of fractal random walks, each with a unique fractal dimension, correlated the laboratory dispersion data and predicted its dependence on column order (non-reciprocity). The fractal dimensions computed from the random walks were dependent upon particle diameter and advective velocity, as shown in Table 4. Although the streamlines were constructed in two dimensions, the dispersion coefficients were independent of the y-direction; i.e., it is a one-dimensional dispersion model, and only the spread in the x-direction was analyzed.

The non-reciprocity of the heterogeneous system demonstrates the need for a description of a porous medium throughout all of a tracer flow path. A heterogeneous system cannot be assigned an average value of dispersivity which can be used to predict dispersion anywhere within the medium. The fractal dimensions computed for columns 123 and 321 were similar in magnitude, but not precisely equal. Therefore, a multifractal walk cannot be completely represented by one fractal dimension.

Although the random walks which were constructed in this study are referred to as fractal walks, in the strict sense they are not fractal curves. A fractal is an object which is self-similar over all range of scales. In reality, of course, a true fractal cannot exist. This is where the notion of a fractal cutoff is employed, represented by ε as in (4.2). The lower fractal cutoff for the walks discussed here is somewhere on the order of a pore diameter. Additionally, there must be an upper fractal cutoff which is somewhat shorter than the length of the system. Hence, the random walks are considered fractal over some discrete range of scales, but simply referred to as fractal. By (4.3), the fractal dimension can be computed without specifying the value of the fractal cutoff.

Future work will concentrate on the scaling of dispersion in porous media using a multifractal model, and the development of a general non-local theory of dispersion.

REFERENCES

[1] Cala M. A., Greenkorn R. A. Velocity effects on dispersion in porous media with a single heterogeneity. *Water Resources Research*, 22(6):919–926, 1986.

[2] Cushman J. H. On unifying the concepts of scale, instrumentation, and stochastics in the development of multiphase transport theory. *Water Resources Research*, 20(11):1668–1676, 1984.

[3] Cushman J. H. Development of stochastic partial differential equations for subsurface hydrology. *Stoch. Hydrol. Hydrau.*, 1:241–262, 1987.

[4] Dagan G. Solute transport in heterogeneous porous formations. *J. Fluid Mech.*, pages 145–151, 1984.

[5] Dagan G. Transport in heterogeneous porous formations: spatial moments, ergodicity, and effective dispersion. *Water Resources Research*, 26(6):1281–1290, 1990.

[6] Gelhar L. W., Montoglou A., Welty C., Rehfeldt K. R. A review of field scale physical solute transport processes in saturated and unsaturated porous media. Technical Report EEPRI EA-4190 Project 2485-5, Electric Power Research Institute, 1985.

[7] Greenkorn R. A. *Flow Phenomena in Porous Media.* Marcel Dekker, Inc., New York, 1983.

[8] Greenkorn R. A., Kessler D. P. Dispersion in heterogeneous non–uniform anisotropic porous media. *Industrial and Engineering Chemistry*, 61(9):14–32, 1969.

[9] Hewett T. A. Fractal distribution of reservoir heterogeneity and their influence on fluid transport. Sixty-first Annual Technical Conference and Exhibition of the Society of Petroleum Engineers, New Orleans, October 5-8, 1986.

[10] Koch D. L., Brady J. F. A non–local description of advection–diffusion with application to dispersion on porous media. *J. Fluid Mech.*, 180:387–403, 1987.

[11] Koch D. L., Brady J. F. Anomalous diffusion in heterogeneous porous media. *Phys. Fluids*, 31(5):965–973, 1988.

[12] Neuman S. P. Universal scaling of hydraulic conductivities and dispersivities in geologic media. *Water Resources Research*, 26(8):1749–1758, 1990.

[13] Ogata A., Banks R. B. A solution to the differential equation of longitudinal dispersion in porous media. Geological Survey Professional Paper 411-A, U.S. Government Printing Office, Washington, D.C., 1961.

[14] Smith L., Schwartz F. W. Mass transport I: a stochastic analysis of macroscopic dispersion. *Water Resources Research*, 16(2):303–313, 1980.

[15] Sternberg S. P. K. An experimental investigation of dispersion in layered heterogeneous porous media. Master's thesis, Purdue University, West Lafayette, IN, 1992.

[16] Sudicky E. A., Gillham R. W., Frind E. O. Experimental investigation of solute transport in stratified porous media, I: the non–reactive case. *Water Resources Research*, 21(7):1035–1042, 1985.

[17] Wheatcraft S. W., Tyler S. W. An explanation of scale dependent dispersivity in heterogeneous aquifers using concepts of fractal geometry. *Water Resources Research*, 24:566–578, 1980.

International Series of Numerical Mathematics, Vol. 114, © 1993 Birkhäuser Verlag Basel

Homogenization in a Perforated Domain Including a Thin Full Interlayer

Alain Bourgeat* Roland Tapiéro†

Abstract. We study the behavior of the solution of a model elliptic problem, the Poisson equation $-\Delta u^\varepsilon = f$, on a domain D_ε including many tiny holes. These holes are periodically distributed with a period diameter equal to ε except in a thin, unperforated layer of thickness η crossing the domain. We show that, if ε, η tend to zero and the ratio η/ε tends to zero or a constant, the behavior of the limit solution is just as if there were no unperforated layer and only a whole, periodically perforated domain. If $\varepsilon < \eta \leq \varepsilon^{\frac{2}{3}}$, the layer has an influence only on itself but not on the perforated parts. If $\eta > \varepsilon^{\frac{2}{3}}$, the asymptotic behavior of the solution is different everywhere in D_ε from the classical solution of Lions [10] in perforated domains. Proofs are given using energy estimates in Sobolev spaces and the framework of homogenization theory; limits can be found by the classical techniques using "test functions" and "two-scale convergence".

1 INTRODUCTION

The aim of this work is to study the limit behavior of some physical processes governed by elliptic partial differential equations of second order in a periodically perforated domain containing a thin layer with no holes. Homogenization has been applied to the study of perforated materials for a long time; see, e.g., Allaire and Murat [2], Cioranescu and Saint Jean Paulin [6], Lévy [9], Lions [10], etc. Many papers on this topic have been devoted to fluid flow in a porous medium; see Arbogast *et al.* [3], Conca [7], Mikelić and Aganović [11], Chapter 7 of Sanchez-Palencia [16], etc. The special question of a medium containing a thin layer with properties different from those of the rest of the material has been the subject of many studies previously; see Ciarlet *et al.* [5], Panasenko [14], Pham Huy and Sanchez-Palencia [15], Chapter 13 of Sanchez-Palencia [16], etc. The authors of this paper have already studied the problem of the Stokes equations but with a supplementary hypothesis on the common trace of the fluid velocity at the boundary of the porous medium and the layer which is, in this case, a fissure through which the fluid flows freely (see Bourgeat *et al.* [4]). We found (and it is also the case when no particular assumption on the common trace is made) that there is a "critical thickness", $\eta = \mathcal{O}(\varepsilon^{\frac{2}{3}})$, for the fissure or layer compared to diameter ε of the period of the perforated domain. It is the widest possible thickness for which the limit behavior in the perforated part is the same as in the usual case of a wholly perforated domain.

We first establish *a priori* estimates in the framework of Sobolev spaces and variational formulations. To find these estimates and then the order of the limits, we

*URA CNRS 0740, Equipe d'Analyse Numérique, Université de Saint-Etienne, 23, Rue du Dr.Paul Michelon, 42023 Saint Etienne Cedex 2, France

†URA CNRS 0740, Laboratoire d'Analyse Numérique, Université Claude Bernard Lyon I, 43, Boulevard du 11 Novembre, 69622 Villeurbanne Cedex, France

use a specific Poincaré-Friedrichs inequality for this kind of domain. To find the limit equations and the above-mentioned properties of their solutions, we use the theory developed by Allaire [1] and Nguesteng [12] of "two-scale convergence", which has proved to be very useful in homogenization theory. We introduce a rescaling in the thin layer which shows that, in the critical case, the limit solution in this layer when ε and η tend to zero is related to the limit in the perforated part by the jump of the normal derivative across the layer. This result is to be compared with the case of a non-perforated domain studied by Pham Huy and Sanchez-Palencia [15]. Unfortunately, this relation is established only in a weak sense (weak convergence in $H^{-1/2}$). A boundary layer in the sense of singular perturbations appears at the junction of the thin layer and the perforated part because of different scales and orders of convergence.

In the following, weak convergence will be denoted by $\rightharpoonup$ and two-scale convergence by — two-scale $\rightharpoonup$.

2 STATEMENT OF THE PROBLEM

Let D be a bounded, connected domain of R^N (the relevant physical cases are $N = 2$ and 3), with a regular boundary ∂D divided into two nonempty parts Γ_1 and Γ_2 (Figure 1). If Σ denotes the intersection of D with the plane $x_N = 0$, then $\Omega = D \backslash \Sigma$ is split into two regions: Ω^+ and Ω^- corresponding to $x_N > 0$ and $x_N < 0$. In order to simplify, let us suppose that Γ_2 contains a cylindrical part, orthogonal to Σ and of the form $\partial\Sigma \times (-\eta_0, +\eta_0) \subset \partial D$, $\eta_0 > 0$. For any small fixed η such that $0 < \eta < \eta_0$, let us consider the thin layer $I_\varepsilon = \Sigma \times (-\eta, +\eta)$, with the cylindrical part of D being bounded at the bottom and top by Σ_ε, the union of $\Sigma_\varepsilon^+ = \Sigma \times \{+\eta\}$ and $\Sigma_\varepsilon^- = \Sigma \times \{-\eta\}$, where ε is the diameter of the period of the perforated domain defined below; the two small parameters ε and η will be related to be defined later (η will be a positive power of ε). For $|x_N| > \eta$ and fixed positive ε, the domain $D \backslash I_\varepsilon$ is assumed to be perforated periodically by a large number of isolated "tiny holes"; this gives a subdomain Ω_ε of $D \backslash I_\varepsilon$ which is the union of two connected parts Ω_ε^+ and Ω_ε^- obtained by intersection with replication starting from Σ_ε^+ and Σ_ε^- of the period εY^*, where Y^*, the generic cell, is a subset of the unit cube $Y = (0,1)^N$ (Figure 2). Let $S_\varepsilon = \partial D_\varepsilon \backslash \partial D$ denote the boundary of the holes.

Let $|.|_\Omega$ denote the usual norm in $L^2(\Omega)$, the Hilbert space of square-integrable functions defined on an open set $\Omega \subset R^N$. Let f be a function in $L^2(D)$ such that

$$|f|_{I_\varepsilon} \leq C\eta^{1/2}. \tag{2.1}$$

For instance, this will happen if f is essentially bounded in D or (and this is the assumption made) if f is *independent* of x_N in the region $\Sigma \times (-\eta_o, +\eta_o)$.

We consider the following problem:

Problem (P_ε). *Find a function u^ε defined in D_ε, satisfying the equations*

$$\begin{aligned} -\Delta u^\varepsilon &= f \text{ in } \Omega_\varepsilon, && (2.2a)\\ u^\varepsilon &= 0 \text{ on } \Gamma_1 \cup S_\varepsilon, && (2.2b)\\ \frac{\partial u^\varepsilon}{\partial n} &= 0 \text{ on } \Gamma_2. && (2.2c) \end{aligned}$$

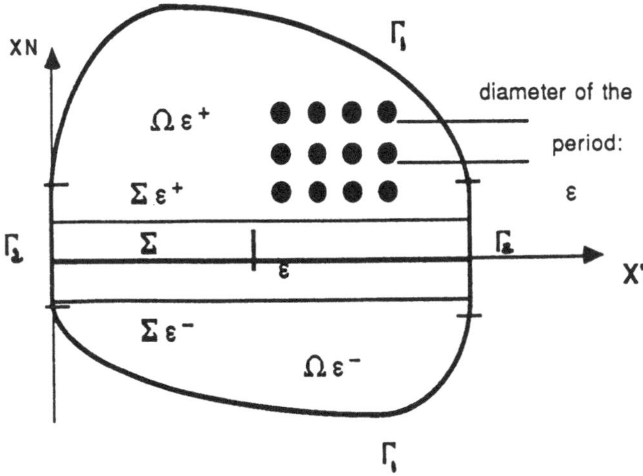

Figure 1: Domain $D_\varepsilon = I_\varepsilon \cup \Sigma_\varepsilon \cup \Omega_\varepsilon$

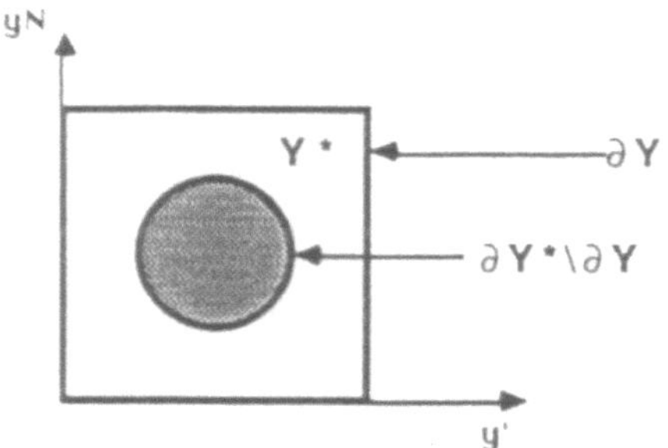

Figure 2: The generic cell Y

3 GLOBAL *a priori* ESTIMATES

Consider spaces

$$\begin{aligned} V(D_\varepsilon) &= \{V \in H^1(D_\varepsilon); v|_{S_\varepsilon} = 0, v|_{\Gamma_1} = 0\}, \\ V(D) &= \{v \in H^1(D); v|_{\Gamma_1} = 0\}, \end{aligned}$$

which are Hilbert spaces for the norms $|\nabla v|_{D_\varepsilon}$ and $|\nabla v|_D$. Note that $v \in V(D_\varepsilon)$ extended by zero in the holes is also a function of $V(D)$.

A "weak resolution" of Problem P_ε is given by finding a function $u^\varepsilon \in V(D_\varepsilon)$ such that, for any $v \in V(D_\varepsilon)$,

$$\int_{D_\varepsilon} \nabla u^\varepsilon \cdot \nabla v dx = \int_{D_\varepsilon} f v \, dx. \tag{3.1}$$

Assume that the hole $Y\backslash Y^*$ in the generic cell is such that the Friedrichs inequality

$$|v|_{Y^*} \leq C|\nabla v|_{Y^*}, \quad \forall v \in H^1(Y^*) \text{ such that } v|_{\partial Y^* \backslash \partial Y} = 0,$$

holds. This is the case if, for instance, the projection of $Y\backslash Y^*$ on a coordinate hyperplane contains an $N-1$ dimensional ball, as proved by Oleinik *et al.* [13] (see also Kondratiev [8]). Then, we have the following specific Poincaré-Friedrichs Inequalities for $V(D_\varepsilon)$.

Proposition 3.1 *For every* $v \in V(D_\varepsilon)$,

$$|v|_{\Omega_\varepsilon} \leq C\varepsilon|\nabla v|_{\Omega_\varepsilon}, \tag{3.2a}$$

$$|v|_{I_\varepsilon} \leq C\left(\sqrt{\eta\varepsilon}|\nabla v|_{\Omega_\varepsilon} + \eta|\nabla v|_{I_\varepsilon}\right), \tag{3.2b}$$

$$|v|_{D_\varepsilon} \leq C(\eta+\varepsilon)|\nabla v|_{D_\varepsilon}. \tag{3.2c}$$

Proof: The inequality (3.2a) is just the Poincaré-Friedrichs Inequality in a perforated domain for a function vanishing on the boundaries of the holes (see Tartar [17]).

Next, consider a point $(x', x_N = \alpha) \in I_\varepsilon$, $x' = (x_1 \ldots x_{N-1}) \in \Sigma$. A dense subspace of $V(D_\varepsilon)$ is given by functions v such that $v \in C^\infty(R^N)$ with compact support in $D_\varepsilon \cup \Gamma_2$, for which

$$v(x', \alpha) = v(x', -\eta) + \int_{-\eta}^{\alpha} \frac{\partial v}{\partial x_N}(x', \xi) d\xi,$$

so that

$$|v(x', \alpha)|^2 \leq 2|v(x', -\eta)|^2 + 4\eta \int_{-\eta}^{\alpha} \left|\frac{\partial v}{\partial x_N}(x', \xi)\right|^2 d\xi.$$

Then, integration over Σ gives

$$\int_\Sigma |v(x', \alpha)|^2 dx' \leq 2\int_\Sigma |v(x', -\eta)|^2 dx' + 4\eta|\nabla v|^2_{I_\varepsilon},$$

which, following integration again from $-\eta$ to $+\eta$, implies that

$$|v|^2_{I_\varepsilon} \leq 4\eta \int_\Sigma |v(x', -\eta)|^2 dx' + 8\eta^2|\nabla v|^2_{I_\varepsilon}. \tag{3.3}$$

Let $\beta_{x'}$ denote the point on S_ε in a one-to-one correspondence with $\alpha_{x'} = (x', -\eta)$; this is possible uniformly in ε by the previous assumption on the hole $Y\backslash Y^*$. Then,

$$v(x', -\eta) = v(\beta_{x'}) + \int_{(\beta_{x'}, \alpha_{x'})} \nabla v \cdot (\alpha_{x'} - \beta_{x'}) d\ell.$$

As $\beta_{x'}$ can be taken in the cell having $\alpha_{x'}$ as a point of its boundary, the distance from $\beta_{x'}$ to $\alpha_{x'}$ is less than ε; since v is equal to zero on S_ε,

$$|v(x', -\eta)|^2 \leq \left|\int_{(\beta_{x'}, \alpha_{x'})} \nabla v \cdot (\alpha_{x'} - \beta_{x'}) d\ell\right|^2 \leq \varepsilon \int_{(\beta_{x'}, \alpha_{x'})} |\nabla v \cdot (\alpha_{x'} - \beta_{x'})|^2 d\ell.$$

If C_ε represents the first layer of cells close to Σ_ε^-, the assumption on Y and integration over Σ give

$$\int_\Sigma |v(x', -\eta)|^2 dx' \leq C\varepsilon \int_{C_\varepsilon} |\nabla v|^2 dx \leq C\varepsilon|\nabla v|^2_{\Omega_\varepsilon}.$$

Then, (3.3) and density in $V(D_\varepsilon)$ imply (3.2b); (3.2a) and (3.2b) give (3.2c). ■

Proposition 3.2 *For every small $\varepsilon > 0$, the unique solution $u^\varepsilon \in V(D_\varepsilon)$ of P_ε in its variational form (3.1) satisfies the* a priori *estimate*

$$|\nabla u^\varepsilon|_{D_\varepsilon} \leq C(\eta^{3/2} + \varepsilon). \tag{3.4}$$

Proof: By (2.1), (3.1), and (3.2*a*),

$$|\nabla u^\varepsilon|^2_{I_\varepsilon} + |\nabla u^\varepsilon|^2_{\Omega_\varepsilon} = \int_{I_\varepsilon} f u^\varepsilon dx + \int_{\Omega_\varepsilon} f u^\varepsilon dx \leq C\left(\eta^{1/2}|u^\varepsilon|_{I_\varepsilon} + \varepsilon|\nabla u^\varepsilon|_{\Omega_\varepsilon}\right).$$

Then, by (3.2*b*),

$$|\nabla u^\varepsilon|^2_{I_\varepsilon} + |\nabla u^\varepsilon|^2_{\Omega_\varepsilon} \leq C\{\eta^{3/2}|\nabla u^\varepsilon|_{I_\varepsilon} + \varepsilon(\eta\varepsilon^{1/2} + \varepsilon)|\nabla u^\varepsilon|_{\Omega_\varepsilon}\}.$$

Therefore,

$$|\nabla u^\varepsilon|^2_{I_\varepsilon} + |\nabla u^\varepsilon|^2_{\Omega_\varepsilon} \leq C\left(\eta^{3/2} + \eta\varepsilon^{1/2} + \varepsilon\right)|\nabla u^\varepsilon|_{D_\varepsilon}.$$

Since $\eta\varepsilon^{1/2} < \eta^{3/2}$ if $\varepsilon < \eta$ and $\eta\varepsilon^{1/2} < \eta^{3/2} < \varepsilon$ if not, the term $\eta\varepsilon^{1/2}$ can be dropped. ∎

The estimate (3.4) means that, for η and ε tending to zero, the extension by zero of $u^\varepsilon/(\eta^{3/2} + \varepsilon)$ is bounded in $V(D)$, which is, for $\eta \leq \varepsilon^{3/2}$, the usual situation in the case of a wholly perforated domain. But, due to the inequalities (3.2*a*) and (3.2*b*), for L^2 estimates in the thin interlayer and in the perforated part, we have to consider three different cases, listed from the worst to the best estimate, in the following proposition.

Proposition 3.3 *The solution $u^\varepsilon \in V(D_\varepsilon)$ of Problem P_ε satisfies the following inequalities:*

a) If $\eta > \varepsilon^{\frac{2}{3}}$,

$$|\nabla u^\varepsilon|_{D_\varepsilon} \leq C\eta^{3/2}, \tag{3.5a}$$
$$|u^\varepsilon|_{I_\varepsilon}\, (or |u^\varepsilon|_{D_\varepsilon}) \leq C\eta^{5/2}, \tag{3.5b}$$
$$|u^\varepsilon|_{\Omega_\varepsilon} \leq C\varepsilon\eta^{3/2}; \tag{3.5c}$$

b) If $\varepsilon < \eta \leq \varepsilon^{\frac{2}{3}}$,

$$|\nabla u^\varepsilon|_{D_\varepsilon} \leq C\varepsilon, \tag{3.6a}$$
$$|u^\varepsilon|_{I_\varepsilon}\, (or |u^\varepsilon|_{D_\varepsilon}) \leq C\eta\varepsilon, \tag{3.6b}$$
$$|u^\varepsilon|_{\Omega_\varepsilon} \leq C\varepsilon^2; \tag{3.6c}$$

c) If $\eta \leq \varepsilon$,

$$|\nabla u^\varepsilon|_{D_\varepsilon} \leq C\varepsilon, \tag{3.7a}$$
$$|u^\varepsilon|_{\Omega_\varepsilon}\, (or |u^\varepsilon|_{D_\varepsilon}) \leq C\varepsilon^2, \tag{3.7b}$$
$$|u^\varepsilon|_{I_\varepsilon} \leq C\eta^{1/2}\varepsilon^{3/2}. \tag{3.7c}$$

Proof: The estimates (3.5a) and (3.6a) follow from Proposition 3.2; (3.2a) and (3.4) imply the estimates on Ω_ε; i.e., (3.5c), (3.6c), and part of (3.7b). From (3.2b),

$$|u^\varepsilon|_{I_\varepsilon} \le C\left(\eta + \eta^{1/2}\varepsilon^{1/2}\right)\left(\eta^{3/2} + \varepsilon\right) \le C\left(\eta^{5/2} + \eta\varepsilon + \eta^{1/2}\varepsilon^{3/2}\right),$$

since $\eta^2\varepsilon^{1/2} < \eta^{5/2}$ for $\eta > \varepsilon$ and $\eta^2\varepsilon^{1/2} \le \eta^2\varepsilon^{3/2} < \eta^{1/2}\varepsilon^{3/2}$ for $\eta \le \varepsilon$. From (3.2c),

$$|u^\varepsilon|_{I_\varepsilon} \le C(\eta + \varepsilon)\left(\eta^{3/2} + \varepsilon\right) \le C\left(\eta^{5/2} + \eta\varepsilon + \varepsilon^2\right). \tag{3.8}$$

Since $\eta^{1/2}\varepsilon^{3/2} < \varepsilon^2$ for $\eta \le \varepsilon$,

$$|u^\varepsilon|_{I_\varepsilon} \le C\left(\eta^{5/2} + \eta\varepsilon + \eta^{1/2}\varepsilon^{3/2}\right); \tag{3.9}$$

(3.8) and (3.9) give (3.5b), (3.6b), (3.7c), and the other part of (3.7b). ∎

Remark. In case a), the thin interlayer I_ε has an influence on the behavior of the solution in the perforated part Ω_ε, since the estimate (3.5c) is worse than ε^2, which is the classical case of the wholly perforated domain as seen in Lions [10]. In case b), we get the classical ε^2-estimate in Ω^ε, but it is yet worse in I_ε. Finally, in case c), the thin interlayer I_ε no longer has an influence on the L^2-estimates, which are those one would find if D_ε were entirely perforated.

4 ESTIMATES OF THE AVERAGED AND RESCALED SOLUTION IN THE LAYER

The estimate (3.9) is on a variable domain I_ε with a Lebesgue measure tending to zero as ε tends to zero. In order to stay in a fixed domain, two operators can be used.

First, Pham Huy and Sanchez-Palencia [15] introduced the "Average in the Layer" operator, which maps functions defined in I_ε into functions defined on Σ by the relation

$$\overline{v}\,(x') = \frac{1}{2\eta}\int_{-\eta}^{+\eta} v\,(x', x_N)dx_N,$$

where $x = (x', x_N) \in I_\varepsilon$ when $x' = (x_1 \ldots x_{N-1}) \in \Sigma$. This operator commutes with

$$\nabla_{x'} = \left(\frac{\partial}{\partial x_1}, \cdots, \frac{\partial}{\partial x_{N-1}}\right)^t,$$

and

$$\frac{\partial \overline{v}}{\partial x_N} = \frac{1}{2\eta}(v(x', \eta) - (v(x', -\eta)), \qquad \forall v \in H^1(I_\varepsilon).$$

This operator is linear and bounded from $L^2(I_\varepsilon)$ into $L^2(\Sigma)$ and from $H^1(I_\varepsilon)$ into $H^1(\Sigma)$; moreover, by Cauchy-Schwarz,

$$|\overline{v}|_\Sigma \le (2\eta)^{1/2}|v|_{I_\varepsilon}. \tag{4.1}$$

A more precise tool than the average above was defined by Ciarlet *et al.* [5] to rescale the problem on I_ε to obtain the fixed domain $I = I_1$. For a function v defined on I_ε, let

$$v(\varepsilon)(x', z) = v(x', x_N = \eta z).$$

As a result of this change of variable,

$$|v(\varepsilon)|_I = \eta^{1/2}|v|_{I_\varepsilon}, \quad \forall L^2(I_\varepsilon); \tag{4.2}$$

for functions in $H^1(I_\varepsilon)$,

$$|\nabla_{x'} v(\varepsilon)|_I = \eta^{1/2}|\nabla_{x'}|_{I_\varepsilon}, \tag{4.3a}$$

$$\left|\frac{\partial v(\varepsilon)}{\partial z}\right|_I = \eta^{1/2}\left|\frac{\partial v}{\partial x_N}\right|_{I_\varepsilon}. \tag{4.3b}$$

Proposition 4.1 *The average in the thickness of the thin interlayer, $\overline{u}$, and the rescaled function, $u(\varepsilon)$, of the solution, u^ε, of P_ε satisfy the following estimates :*

$$\sqrt{2}|\overline{u}^\varepsilon|_\Sigma \le |u(\varepsilon)|_I \le C\left(\eta^2 + \eta^{1/2}\varepsilon + \varepsilon^{3/2}\right), \tag{4.4a}$$

$$\sqrt{2}|\nabla_{x'}\overline{u}^\varepsilon|_\Sigma \le |\nabla_{x'} u(\varepsilon)|_I \le C\left(\eta^2 + \eta^{1/2}\varepsilon\right), \tag{4.4b}$$

$$\frac{1}{\sqrt{2}}|u^\varepsilon(x', \eta) - u^\varepsilon(x', -\eta)|_\Sigma \le \left|\frac{\partial u(\varepsilon)}{\partial z}\right|_I \le C\left(\eta^2 + \eta^{1/2}\varepsilon\right). \tag{4.4c}$$

Proof: The bounds (4.4*a*) and (4.4*b*) come from the equalities (4.2) and (4.3*a*) and the inequalities (4.1), (3.9), and (3.4). Then, (4.4*c*) is obtained by applying Cauchy-Schwarz and (3.4) in combination with (4.3*b*). ■

Remarks. If $\eta > \varepsilon$ (thereby excluding case c) of Proposition 3.3 for which the thin layer has no influence), then

$$|u(\varepsilon)|_I \le C\left(\eta^2 + \eta^{1/2}\varepsilon\right),$$

$$\left|\frac{\partial u(\varepsilon)}{\partial z}\right|_I \le C\left(\eta^2 + \eta^{1/2}\varepsilon\right),$$

$$|\nabla_{x'} u(\varepsilon)|_I \le C\left(\eta + \eta^{-1/2}\varepsilon\right).$$

These estimates indicate a ratio η for the L^2 estimate of $u(\varepsilon)$ and its gradient with respect to x'. Moreover, *the best estimate is obtained for* $\eta = \mathcal{O}(\varepsilon^{\frac{2}{3}})$, the maximum value of the thickness for which we still have the classical $\mathcal{O}(\varepsilon^2)$-behavior in the perforated part for the L^2-norm of u^ε and $\mathcal{O}(\varepsilon)$ for its H^1-norm.

5 TRACE ESTIMATES ON Σ

Proposition 5.1 *The traces on Σ_ε and on Σ of the solution $u^\varepsilon \in V(D_\varepsilon)$ of P_ε satisfy the following estimates:*

$$|u^\varepsilon|_{\Sigma_\varepsilon}|_\Sigma \le C\varepsilon^{1/2}\left(\eta^{3/2} + \varepsilon\right), \tag{5.1a}$$

$$|u^\varepsilon|_\Sigma|_\Sigma \le C\left(\eta^2 + \eta^{1/2}\varepsilon + \varepsilon^{3/2}\right). \tag{5.1b}$$

Proof: As a function in $H^1(D)$, u^ε has traces $u^\varepsilon|_{\Sigma_\varepsilon}$ and $u^\varepsilon|_\Sigma$ in $L^2(\Sigma)$. As in Proposition 3.1, introduce $\beta_{x'}$, the point on S_ε corresponding to $\alpha_{x'} = (x', \eta)$. Then,

$$|u^\varepsilon|_{\Sigma_\varepsilon^+}|_\Sigma^2 = \int_\Sigma (u^\varepsilon(x', \eta))^2 dx' \le 2\int_\Sigma \left(\int_{(\beta_{x'}, \alpha_{x'})} \nabla u^\varepsilon \cdot (\alpha_{x'} - \beta_{x'}) d\ell\right)^2 dx',$$

so that, by Cauchy-Schwarz,

$$|u^{\varepsilon}{}_{|\Sigma_\varepsilon}|^2_{\Sigma} \leq C\varepsilon|\nabla u^{\varepsilon}|^2_{\Omega\varepsilon}, \tag{5.2}$$

which, with (3.4), gives (5.1a). Otherwise,

$$|u^{\varepsilon}|_{\Sigma}|^2_{\Sigma} = \int_{\Sigma}(u^{\varepsilon}(x',0))^2dx' \leq 2\int_{\Sigma}\left(\int_0^{\eta}\frac{\partial u^{\varepsilon}}{\partial x_N}dx_N\right)^2 dx';$$

thus, (5.2) and Cauchy-Schwarz imply that

$$|u^{\varepsilon}|_{\Sigma}|^2_{\Sigma} \leq C\left(\varepsilon|\nabla u^{\varepsilon}|^2_{\Omega_\varepsilon} + \eta|\nabla u^{\varepsilon}|^2_{I_\varepsilon}\right).$$

Then, (5.1b) follows from (3.4) and the argument used to prove (4.4a). ■

Remark. The same *a priori* estimate for the trace on Σ holds as held for the average in the thickness.

6 A CONVERGENCE RESULT

Let $H^1_\sharp(Y^*)$ denote the space of functions of $H^1(Y^*)$ which vanish on $\partial Y^*\backslash\partial Y$ and are extended by zero on the void part $Y\backslash Y^*$ and have equal traces on the opposite faces of Y and are periodically repeated on the whole space R^N. Let $w \in H^1_\sharp(Y^*)$ be the solution of the variational problem

$$\int_{Y^*}\nabla_y w(y)\cdot\nabla_y v(y)\,dy = \int_{Y^*} v(y)\,dy, \qquad \forall v \in H^1_\sharp(Y^*);$$

w is also the solution of the corresponding strongly formulated problem:

$$\begin{aligned} &-\Delta_y w(y) = 1 \quad \text{in} \quad Y, \\ &w|_{\partial Y^*\backslash\partial Y} = 0, \\ &w \quad \text{is} \quad Y\text{-periodic.} \end{aligned}$$

Finally, set

$$\widetilde{w} = \int_{Y^*} w\,(y)dy.$$

Then, we have a global convergence result.

Theorem 6.1 *Extend u_ε by zero in the holes. Then, for $\eta \leq \varepsilon^{\frac{2}{3}}$,*

$$\varepsilon^{-2}u^{\varepsilon} \rightharpoonup f(x)\widetilde{w} \text{ in } L^2(D) \text{ as } \varepsilon \to 0.$$

For $\varepsilon^{\frac{2}{3}} < \eta$,

$$(\varepsilon\eta^{3/2})^{-1}u^{\varepsilon} \rightharpoonup 0 \text{ in } L^2(D) \text{ as } \eta \to 0.$$

For $\eta \geq \varepsilon^{\frac{2}{3}}$, the jump of the normal derivative across the thin layer is given by

$$\frac{1}{2\eta}\left[\frac{\partial u^{\varepsilon}}{\partial x_N}\right]_{I_\varepsilon} = \frac{1}{2\eta}\left(\frac{\partial u^{\varepsilon}}{\partial x_N}(x',\eta) - \frac{\partial u^{\varepsilon}}{\partial x_N}(x'-\eta)\right);$$

the jump converges weakly in $H^{-1/2}(\Sigma)$ to $-f(x')$.

The proof will be obtained through a sequence of lemmata using the "two-scale convergence" introduced by Nguesteng [12] and developed by Allaire [1].

Lemma 6.1 *For* $\eta \leq \varepsilon$,

$$\begin{aligned} \varepsilon^{-2}u^\varepsilon \quad &- \textit{two-scale} \rightarrow u(x,y) \in \ L^2(D; H^1_\sharp(Y^*)), \\ \varepsilon^{-1}\nabla u^\varepsilon \quad &- \textit{two-scale} \rightarrow \nabla_y u(x,y), \end{aligned}$$

where $u(x,y) = f(x)w(y)$ *and is the solution of the variational homogenized problem*

$$\int_D \int_{Y^*} \nabla_y u(x,y) \cdot \nabla_y \varphi(x,y) dx\, dy = \int_D \int_{Y^*} f(x)\varphi(x,y) dx\, dy, \quad \forall \varphi \in L^2(D; H^1_\sharp(Y^*)).$$

Proof: The two-scale convergence result and existence of $u(x,y)$ follow from (3.6a) and (3.7b); see Allaire [1]. Then, take the test function in (3.1) to be $\varphi(x, x/\varepsilon)$, where $\varphi(x,y) \in C_0^\infty(D; C_\sharp^\infty(Y^*))$ and vanishes for $y \in Y \backslash Y^*$, to obtain the relation

$$\int_D \varepsilon^{-1} \nabla u^\varepsilon \left(\nabla_y \varphi(x, x/\varepsilon) + \varepsilon \nabla_x \varphi(x, x/\varepsilon)\right) dx = \int_D f(x)\varphi(x, x/\varepsilon) dx.$$

Lemma 6.1 is proved by passing to the two-scale limit and noting the density of the test functions used. ■

Lemma 6.2 *For* $\varepsilon < \eta \leq \varepsilon^{\frac{2}{3}}$,

$$\begin{aligned} \varepsilon^{-2}u^\varepsilon|_{\Omega_\varepsilon^+} \quad &- \textit{two-scale} \rightarrow f(x)w(y) \ \textit{in} \ L^2(\Omega^+; H^1_\sharp(Y^*)); \\ \varepsilon^{-1}\nabla u^\varepsilon|_{\Omega_\varepsilon^+} \quad &- \textit{two-scale} \rightarrow f(x)\nabla_y w(y) \ \textit{in} \ L^2(\Omega^+; L^2_\sharp(Y^*)). \end{aligned}$$

Analogous results hold on Ω_ε^-.

Proof: The proof is similar to the proof of Lemma 6.1, except that we take as test function in (3.1) the function $\varphi(x, x/\varepsilon)$, where $\varphi(x,y) \in C_0^\infty(\Omega^+; C_\sharp^\infty(Y^*))$ (respectively, $\varphi(x,y) \in C_0^\infty(\Omega^-; C_\sharp^\infty(Y^*))$). ■

Lemma 6.3 *For* $\eta > \varepsilon^{\frac{2}{3}}$,

$$\begin{aligned} (\varepsilon\eta^{3/2})^{-1}u^\varepsilon|_{\Omega_\varepsilon^+} \quad &- \textit{two-scale} \rightarrow 0 \ \textit{in} \ L^2(\Omega^+; H^1_\sharp(Y^*)); \\ \eta^{-3/2}\nabla u^\varepsilon|_{\Omega_\varepsilon^+} \quad &- \textit{two-scale} \rightarrow 0 \ \textit{in} \ L^2(\Omega^+; H^1_\sharp(Y^*)). \end{aligned}$$

Analogous results hold on Ω^-.

Proof: As consequences of (3.5a) and (3.5c), $u^\varepsilon/\varepsilon\eta^{3/2}$ and $\nabla u^\varepsilon/\eta^{3/2}$ two-scale converge to some $u^0(x,y)$ and $\nabla_y u^0(x,y)$ in Ω^+, respectively. Then, taking the test function in (3.1) to be $\varepsilon\eta^{-3/2}\varphi(x, x/\varepsilon)$, where $\varphi(x,y) \in C_0^\infty(\Omega^+; C_\sharp^\infty(Y^*))$ and vanishes for $y \in Y\backslash Y^*$, gives

$$\int_{\Omega^+} \eta^{-3/2}\nabla u^\varepsilon \cdot (\nabla_y \varphi(x, x/\varepsilon) + \varepsilon \nabla_x \varphi(x, x/\varepsilon)) dx = \varepsilon\eta^{-3/2} \int_{\Omega^+} f(x)\varphi(x, x/\varepsilon)) dx.$$

As $\varepsilon\eta^{-3/2}$ tends to zero, we see, by passing to the two-scale limit that

$$\int_{\Omega^+}\int_{Y^*} \nabla_y u^0(x,y)\cdot\nabla_y\varphi(x,y)\,dx\,dy = 0.$$

Thus, u^0 is a constant with respect to y which is zero because of the homogeneous boundary condition on $\partial Y^*\backslash\partial Y$. The same argument is applicable when Ω^+ is replaced by Ω^-. ■

Proof of Theorem 6.1. The fundamental result of two-scale convergence (Nguesteng [12] and Allaire [1]) is that $\varepsilon^{-2}u^\varepsilon$ converges weakly in $L^2(D)$ to $\int_{Y^*} u(x,y)dy$. Then, Lemmata 6.1 and 6.2 imply the first conclusion of the theorem, since weak convergence in $L^2(\Omega^+\cup\Omega^-)$ is the same as in $L^2(D)$, because the measure of Σ in R^N is zero. Lemma 6.3 gives the second part.

The critical case is $\eta = \mathcal{O}(\varepsilon^{\frac{2}{3}})$, for which (4.4$a$) and (4.4$c$) applied to the rescaled function $u(\varepsilon)$ give the following convergence results:

$$\begin{aligned}\eta^{-2}u(\varepsilon) &\rightharpoonup u(0)(x',z) \text{ in } L^2(I),\\ \eta^{-2\frac{\partial u(\varepsilon)}{\partial z}} &\rightharpoonup \frac{\partial u(0)}{\partial z}(x',z) \text{ in } L^2(I).\end{aligned}$$

Consider P_ε as formulated in I_ε and rescaled in I and test by $v\in H^1(I)$. Then,

$$\begin{aligned}&\int_I \nabla_{x'}u(\varepsilon)\cdot\nabla_{x'}v\,dx'dz + \frac{1}{\eta^2}\int_I \frac{\partial u(\varepsilon)}{\partial z}\frac{\partial v}{\partial z}\,dx'dz\\ &\quad -\frac{1}{\eta^2}\int_\Sigma\left(\frac{\partial u(\varepsilon)}{\partial z}(x',1)v(x',1) - \frac{\partial u(\varepsilon)}{\partial z}(x',-1)v(x',-1)\right)dx'\\ &= \int_I fv\,dx'dz.\end{aligned}$$

Apply (4.4b) and pass to the limit; thus,

$$\begin{aligned}&\int_I \frac{\partial u(0)}{\partial z}\frac{\partial v}{\partial z}\,dx'dz - \lim_{\eta\to 0}\frac{1}{\eta}\int_\Sigma\left(\frac{\partial u^\varepsilon}{\partial x_N}(x',\eta)v(x',1) - \frac{\partial u^\varepsilon}{\partial x_N}(x',-\eta)v(x',-1)\right)dx'\\ &= \int_\Sigma f(x')\left(\int_{-1}^{+1} v(x',z)dz\right)dx'.\end{aligned}$$

If v is taken independent of z, then

$$\lim_{\eta\to 0}\int_\Sigma \frac{1}{\eta}\left[\frac{\partial u^\varepsilon}{\partial x_N}\right]_{I_\varepsilon} v(x')\,dx' = -2\int_\Sigma f(x')v(x')dx',$$

and the final conclusion of the theorem has been demonstrated. ■

7 CONCLUSION

The theorem gives only weak limits. To have more precise results and, in particular, to have a stronger relation between the limit solution in the layer and in the perforated part than the one obtained in the last part of the convergence theorem, one should introduce correctors, since a boundary layer appears between the two media because of the different scales and orders in the estimates.

REFERENCES

[1] Allaire G. Homogenization and two scale convergence. *SIAM J. of Math. Anal.*, 23:1482–1518, 1992.

[2] Allaire G., Murat F. Homogenization of Neuman problem with non-isolated holes. To appear 1992.

[3] Arbogast T., Douglas J. Jr., Hornung U. Derivation of the double porosity model of single phase flow via homogenization theory. *SIAM J. of Math. Anal.*, 21:823–836, 1990.

[4] Bourgeat A., Elamri H., Tapiéro R. Existence d'une taille critique pour une fissure dans un milieu poreux. In *Second Colloque Franco-Chilien de Mathématiques Appliquées*, pages 67–80, Toulouse, 1991. Cépadués Editions. C. Carasso, C. Conca, R. Correa, J. P. Puel, eds.

[5] Ciarlet P. G., Ledret H., Nzwenga R. Modélisation de la jonction entre un corps élastique tridimensionnel et une plaque. *C. R. Acad. Sci., Paris*, Série I, t. 305:55–58, 1987.

[6] Cioranescu D., Saint Jean Paulin J. Homogenization in open sets with holes. *J. of Math. Anal. and Appl.*, 71:590–607, 1979.

[7] Conca C. Étude d'un fluide traversant une paroi perforée, I et II. *J. Math. Pures et Appl.*, 66:1–70, 1987.

[8] Kondratiev V. A. On the solvability of the first boundary value problem for strongly elliptic equations. *Trans. of the Moscow Math. Soc.*, 16:293–318, 1967.

[9] Lévy T. Écoulement d'un fluide dans un milieu poreux fissuré. *C. R. Acad. Sci., Paris*, Série II, t. 306:1413–1417, 1988.

[10] Lions J. L. *Some Methods in the Mathematical Analysis of Systems and Their Control.* Science Press Beijing, Gordon and Breach Science Publishers, Inc., New York, 1981.

[11] Mikelić A., Aganović I. Homogenization in a porous medium under non-homogeneous boundary conditions. *Bolletino U. M. I.*, 7:171–180, 1987.

[12] Nguesteng G. A general convergence result for a functional related to the theory of homogenization. *SIAM J. of Math. Anal.*, 20:608–623, 1989.

[13] Oleinik O. A., Shamaev A. S., Yosifian G. A. Asymptotic expansions of solutions of the Dirichlet problem for elliptic equations in perforated domains. *Matem. Sbornik*, 112(154):588–610, 1980.

[14] Panasenko G. P. Higher order asymptotics of solutions of problems on the contact of periodic structures. *Math. U.S.S.R. Sbornik*, 38:465–494, 1981.

[15] Pham Huy H., Sanchez-Palencia E. Phénomènes de transmission à travers des couches minces de conductivité élevée. *J. of Math. Anal. and Appl.*, 47:284–309, 1974.

[16] Sanchez-Palencia E. *Non Homogeneous Media and Vibration Theory.* Lecture Notes in Physics 127. Springer Verlag, Berlin, 1980.

[17] Tartar L. Incompressible fluid flow in a porous medium — Convergence of the homogenization process. Appendix, Sanchez-Palencia [16], 1980.

International Series of Numerical Mathematics, Vol. 114,

Indirect Determination of Hydraulic Properties of Porous Media

J. R. Cannon* Paul DuChateau†

Abstract. This paper considers an inverse problem associated with modelling unsaturated flow in a porous medium. An unknown coefficient in a typical inverse problem is shown to be uniquely determined in a given equivalence class by a combination of Dirichlet and Neumann data. It is further shown that an approximation to the coefficient can be constructed by an algorithm based on a trace-type functional formulation of the inverse problem.

1 INTRODUCTION

Single-phase and multi-phase flow in an unsaturated porous medium can be modelled by nonlinear partial differential equations in which the coefficients are related to the hydraulic properties of the medium. Treatment of these equations is considerably simplified if we assume that the coefficients are functions of the unknown dependent variable only. In such cases it is often then feasible to experimentally determine the coefficients for the purpose of modelling a specific porous medium. There are two alternative approaches for this determination:

Direct methods - hydraulic properties are measured directly in a (usually) very complicated physical experiment.

Indirect methods - hydraulic properties are obtained by formulating and solving an appropriate inverse problem. Input data for the inverse problem is obtained from relatively simple physical experiments.

Indirect methods can be further classified into history matching schemes and coefficient reconstruction schemes. History-matching schemes characterize the unknown coefficients in terms of a finite number of parameters and then seek to adjust these parameters so that computed output generated by the associated mathematical model matches experimentally measured output in some suitable sense. These methods have the advantage that the inverse problem can be recast as an optimization problem for which efficient computational algorithms are readily available. The disadvantage to history-matching schemes lies in the fact that a solution to the optimization problem is not necessarily a solution to the inverse problem.

Here we will illustrate a coefficient reconstruction scheme in which the unknown coefficient is expressed directly in terms of overspecified data. This leads to an initial boundary value problem involving a trace-type functional partial differential equation and it is not hard to show that any solution of this problem is also a solution of the

*Department of Mathematics, Lamar University, P. O. Box 10047, Beaumont, TX 77710

†Department of Mathematics, Colorado State University, Fort Collins, CO 80523

inverse problem [6]. Although we cannot yet prove existence of a solution to this problem, we can show that the overspecified data uniquely determines the unknown coefficient within an appropriate equivalence class. If the solution is assumed to exist, then we can demonstrate a rather efficient algorithm for constructing an approximation to the solution.

We will begin by considering the so-called direct problem. We derive needed properties of the solution of this initial boundary value problem where the unknown coefficient is assumed to be known and no extra boundary data is given. We then proceed to formulate the inverse problem and prove a uniqueness result. Finally, we describe an algorithm for approximating a solution to the inverse problem and illustrate the algorithm with some numerical experiments.

2 THE DIRECT INITIAL BOUNDARY VALUE PROBLEM

We consider an initial boundary value problem for a nonlinear diffusion equation. The diffusivity coefficient will be chosen from the following class of admissible coefficients.

Definition 2.1 *The coefficient $a = a(u)$ belongs to the class A if and only if*

i) $a \in C^1(-\infty, \infty)$,

ii) *for fixed, positive constants a_0 and $a_1, a_0 \le a(u) \le a_1$ for all u.*

It is convenient to define another class of admissible data.

Definition 2.2 *If, for some $T > 0$,*

i) $f \in C^1[0,T]$ *with* $f'(t) > 0$ *for* $0 < t < T$,

ii) $f(0) = f'(0) = 0$,

f belongs to the class D of admissible data.

Then, for $a \in A$ and $f \in D$, we consider the initial boundary value problem

$$\begin{aligned} \partial_t u(x,t) &= \partial_x(a(u)\partial_x u(x,t)) \text{ for } 0 < x < 1, \quad 0 < t < T, \\ u(x,0) &= 0 \quad \text{for } 0 < x < 1, \\ u(0,t) &= f(t), \quad u(1,t) = 0, \quad \text{for } 0 < t < T. \end{aligned} \tag{2.1}$$

In Gilding [6], it is shown that, for each $a \in A$ and every $f \in D$, the initial boundary value problem (2.1) has a unique solution $u = u(x,t)$ in $C^{2,1}(Q_T)$, where $Q_T = (0,1) \times (0,T)$, and that the function

$$g(t) := -a(f(t))\partial_x u(0,t) \tag{2.2}$$

belongs to $C^1(0,T)$. We can derive some additional information about the solution $u = u(x,t)$.

Theorem 2.1 *Suppose that $u = u(x,t)$ denotes the unique smooth solution of (2.1) for some $a \in A$ and some $f \in D$. Then,*

$$\partial_t u(x,t) > 0 \quad \textit{for } 0 < x < 1, \quad 0 < t < T, \tag{2.3}$$

and

$$f(t) > u(x,t) > z(x,t) \ \textit{for}\ 0 < x < 1, \quad 0 < t < T.$$

Here, $z = z(x,t)$ denotes the solution of the auxiliary problem

$$\begin{aligned} \partial_t z(x,t) &= a_0 \partial_{xx} z(x,t) + b \partial_x z(x,t) \ \textit{in}\ Q_T, \\ z(x,0) &= 0, \qquad 0 < x < 1, \\ z(0,t) &= f(t), \quad z(1,t) = 0, \quad 0 < t < T, \end{aligned} \tag{2.4a}$$

where a_0 denotes the positive lower bound of a required by Definition 2.1 and b is a sufficiently large positive constant whose size depends on $a(u)$ and $f(t)$.

Proof: If we let

$$v(x,t) = \int_0^{u(x,t)} a(s)\,ds =: A(u(x,t)),$$

then $v = v(x,t)$ solves the problem

$$\begin{aligned} \partial_t v(x,t) &= a(A^{-1}(v)) \partial_{xx} v(x,t) \ \text{in}\ Q_T, \\ v(x,0) &= 0, \quad 0 < x < 1, \\ v(0,t) &= F(t) = A(f(t)), \quad v(1,t) = 0, \quad 0 < t < T. \end{aligned}$$

It is not hard to show that the hypotheses imply that $\partial_t v(x,t) > 0$ in Q_T; see DuChateau, [5]. Then, since $\partial_t v(x,t) = a(u)\partial_t u(x,t)$, (2.3) follows.

It follows from the strong max-min principle applied to (2.1) that

$$f(t) > u(x,t) > 0 \ \text{for}\ 0 < x < 1, \quad 0 < t < T. \tag{2.5}$$

Now, if $z = z(x,t)$ solves (2.4), then it follows that $w(x,t) =: u(x,t) - z(x,t)$ must satisfy

$$\begin{aligned} \partial_t w - \partial_x(a(u)\partial_x w) &= (a(u) - a_0)\partial_{xx} z + (a'(u)\partial_x u - b)\partial_x z \\ w(x,0) &= w(0,t) = w(1,t) = 0. \end{aligned}$$

An application of the minimum principle to this initial boundary value problem leads to the other half of the estimate (2.4), provided we can show that

$$(a(u) - a_0)\partial_{xx} z + (a'(u)\partial_x u - b)\partial_x z > 0 \ \text{in}\ Q_T.$$

Obviously $a(u) - a_0$ is positive, and, for $b > 0$ sufficiently large, $a'(u)\partial_x u - b$ is negative on Q_T since $a'(u)\partial_x u$ is bounded there. In particular, it is sufficient to choose

$$b \geq \sup_{Q_T} \mid a'(u(x,t))\partial_x u(x,t) \mid.$$

It remains to show that $\partial_{xx} z$ is positive and $\partial_x z$ is negative on Q_T. This follows from the observation that, if we let

$$z(x,t) = e^{\alpha t + \beta x} v(x,t), \quad \text{for}\ \alpha = -\frac{b^2}{4a_0} \ \text{and}\ \beta = -\frac{b}{2a_0},$$

then $v = v(x,t)$ is the solution of

$$\begin{aligned} \partial_t v &= a_0 \partial_{xx} v \ \text{ in } Q_T, \\ v(x,0) &= 0, \quad \text{for } 0 < x < 1, \\ v(0,t) &= e^{-\alpha t} f(t), \quad v(1,t) = 0, \quad \text{for } 0 < t < T. \end{aligned}$$

Since $\alpha < 0$, and $f \in D$, it is easy to show that

$$\partial_{xx} v(x,t) > 0,\ \partial_x v(x,t) < 0, \quad \text{and } v(x,t) > 0 \ \text{ for } 0 < x < 1,\ 0 < t < T.$$

Then, since $\beta < 0$, we have

$$\begin{aligned} z(x,t) &= e^{\alpha t + \beta x} v(x,t) > 0, \\ \partial_x z(x,t) &= e^{\alpha t + \beta x} (\partial_x v(x,t) + \beta v(x,t)) < 0, \\ \partial_{xx} z(x,t) &= e^{\alpha t + \beta x} (\partial_{xx} v(x,t) + 2\beta \partial_x v(x,t) + \beta^2 v(x,t)) > 0. \end{aligned}$$

This completes the proof. ∎

Lemma 2.1 *Suppose $u = u(x,t)$ solves (2.1) for $a \in A$ and $f \in D$. Then for each positive T, $g = g(t)$ given by (2.2) satisfies*

$$g(0) = 0 \ \textit{ and } g(t) > 0 \ \textit{ for } 0 < t < T. \tag{2.6}$$

Proof: It follows from (2.5) that for each $x > 0$,

$$\frac{f(t) - z(x,t)}{0 - x} < \frac{f(t) - u(x,t)}{0 - x} < 0, \quad \text{for } 0 < t < T.$$

Letting x tend to zero through positive values leads to

$$\partial_x z(0,t) < \partial_x u(0,t) < 0, \quad \text{for } 0 < t < T.$$

This implies $g(t) > 0$ for $0 < t < T$, and since $\partial_x z(0,t)$ tends to zero as t tends to zero, we get $g(0) = 0$.

In the context of heat conduction the function $g(t)$ defined in (2.2) can be interpreted as the boundary heat flux corresponding to a boundary temperature $f(t)$. In the language of partial differential equations, $f(t)$ and $g(t)$ are referred to as Dirichlet data and Neumann data, respectively. For a fixed $a \in A$, the initial boundary value problem (2.1) induces, via (2.2), a one-to-one correspondence between f and g; we refer to this correspondence as the "Dirichlet to Neumann" mapping and write $g = N[f;a]$. According to Lemma 2.1, the condition (2.6) is necessary in order that $g = N[f;a]$ for $a \in A$ and $f \in D$. There is an additional necessary condition. ∎

Theorem 2.2 *Suppose $u = u(x,t)$ solves (2.1) for $a \in A$ and $f \in D$ and that $g(t)$ is given by (2.2). Then, it follows that*

$$\left(\lim_{t \to 0} \frac{g(t)}{H(t)} \right)^2 = a(f(0)) = a(0), \tag{2.7}$$

where

$$H(t) = \int_0^t \frac{f'(\tau)}{\sqrt{\pi(t-\tau)}} d\tau; \tag{2.8}$$

i.e., the limit exists and it equals $a(0)$.

A proof of this result can be found in Cannon and DuChateau [2, 3]. Summarizing the last two results, we observe that, for $a \in A$ and f in D,

$$\begin{aligned} N[f;a] &\in C^1[0,T], \\ N[f;a](0) &= 0, \text{ and } N[f;a](t) > 0, \text{ for } t > 0, \\ \lim_{t\to 0}[N[f;a](t)/H(t)] &= \mu, \text{ and } \mu^2 = a(0). \end{aligned} \tag{2.9}$$

If $g = g(t)$ satisfies the conditions listed in (2.9), we say that g lies in the range of the Dirichlet to Neumann map and write $g \in \mathcal{R}(N)$.

3 THE INVERSE PROBLEM

Now, we consider an inverse problem associated with the initial boundary value problem (2.1). Let f and g be given, with $f \in D$, and $g \in \mathcal{R}(N)$. Then we seek a pair of functions, $u = u(x,t)$ in $C^{2,1}(Q_T)$ and $a = a(u)$ in A satisfying

$$\begin{aligned} \partial_t u(x,t) &= \partial_x(a(u)\partial_x u(x,t)) \text{ in } Q_T, \\ u(x,0) &= 0, \quad 0 < x < 1, \\ u(0,t) &= f(t), \quad u(1,t) = 0, \quad 0 < t < T, \\ -a(f(t))\partial_x u(0,t) &= g(t), \quad 0 < t < T. \end{aligned} \tag{3.1}$$

We point out that in a physical experiment the data function $f = f(t)$ would be *controlled* in such a way that f belongs to D. Simultaneously, the data function $g = g(t)$ would then be *observed*; the results of the previous section indicate that for $f \in D$, g would be found to be in $\mathcal{R}(N)$. Note that the data functions f and g determine the value $a(0) = a(f(0))$ via (2.7) and (2.8). In addition, the extent of the domain over which $a(u)$ is determined by the inverse problem (3.1) is controlled by $f(t)$. In particular, we shall show that (3.1) defines $a(u)$ on the interval, $0 \le u \le f(T)$.

In Cannon *et al.* [4], it is shown that the coefficient $a(u)$ can be expressed directly in terms of data and eliminated from the nonlinear diffusion equation. The resulting initial boundary value problem is then a so-called *trace-type functional problem.* Existence of a solution to that problem is then equivalent to existence of a solution to the inverse problem.

Before considering the question of existence for a solution to the inverse problem, we must know whether the conditions of (3.1) define the solution uniquely; i.e., we have to know if there can be two different coefficients $a = a(u)$ in A that satisfy the conditions of (3.1). In order to answer that question we must specify what is meant by *different* coefficients. Eventually, we are going to construct a continuous, piecewise linear approximation for the unknown coefficient $a = a(u)$. With that in mind, we define an equivalence relation on the class A of admissible coefficients.

Definition 3.1 *Coefficients a and b in A are said to be* separated by a linear graph *on $[\alpha,\beta]$ if there exists an interval $(c, c+\epsilon) \subset [\alpha,\beta]$ of positive length on which the graph of $a(u)$ can be separated from the graph of $b(u)$ by a straight line segment; i.e., there exists $(c, c+\epsilon) \subset [\alpha,\beta]$ with $\epsilon > 0$, and linear function $\Lambda(u)$ such that*

$\Lambda(u)$ lies strictly between $a(u)$ and $b(u)$ for $c < u < c+\epsilon$,
$\Lambda'(u)$ lies between $a'(u)$ and $b'(u)$ for $c < u < c+\epsilon$.

If $a(u)$ and $b(u)$ are not separated by a linear graph, then they will generate the same piecewise linear (i.e., polygonal) approximation. In this case, $a(u)$ and $b(u)$ are said to be polygonally equivalent *on $[\alpha, \beta]$; this relation will be denoted by $a \approx b$ and defines an equivalence relation on the class A of admissible coefficients.*

We will need an auxiliary result relating to the solution of (2.1) in case the coefficient $a(u)$ is a *linear* function of u. For $f = f(t)$ in D, let α and β denote constants such that $a_0 \leq \alpha + \beta f(T) \leq a_1$. Then, $a(v) = \alpha + \beta v \in A$, and we may consider $v = v(x,t)$, the solution for initial boundary value problem (2.1) corresponding to $f \in D$ and this linear coefficient in A. Now, we suppose that $f(t)$ is such that there exists a T_0, $0 < T_0 \leq T$, depending on α, β and $f(t)$, such that

$$\partial_{xx} v(x,t) \geq 0 \quad \text{on } Q_{T_0}. \tag{3.2}$$

For $\alpha > 0$ and $\beta \leq 0$ it is easy to show that (3.2) holds for all $f \in D$. We conjecture that, even in the case $\beta > 0$, requiring f to be in D is sufficient to imply (3.2). Proving this conjecture is an open problem.

Theorem 3.1 *Suppose that $f = f(t) \in D$ is such that (3.2) holds and that both $a(u)$ and $b(u)$ are in A. Then let $u = u(x,t;a)$ and $v = v(x,t;b)$ denote the corresponding solutions of the initial boundary value problem (2.1). Suppose further that*

$$a(f(t))\partial_x u(0,t) = b(f(t))\partial_x v(0,t) \quad \text{for} \quad 0 \leq t \leq T. \tag{3.3}$$

Then, a and b are polygonally equivalent on $[0, f(T)]$; i.e., if $N[f;a] = N[f;b]$, then $a \approx b$.

Proof: Suppose that $a = a(u)$ and $b = b(u)$ are separated by a linear graph on $[0, f(T)]$. Then for c and $\epsilon > 0$ as in Definition 3.1, let t_0 and t_1 be chosen so that $f(t_0) = c$ and $f(t_1) = c + \epsilon$. In fact, the argument loses no generality if we suppose that $t_0 = 0$, and we have

$$\begin{aligned} a(0) &= b(0) = \Lambda(0) > 0, \\ a(s) &> \Lambda(s) > b(s) \quad \text{for } 0 < s < f(t_1), \\ a'(s) &\geq \Lambda'(s) \geq b'(s) \quad \text{for } 0 < s < f(t_1). \end{aligned} \tag{3.4}$$

Cases different from (3.4) can be handled by slight modifications of the argument which follows. Now (3.3) and (3.4) together imply that

$$\partial_x u(0,t) < \partial_x v(0,t), \qquad \text{for } 0 < t < t_1. \tag{3.5}$$

Let $z(x,t)$ denote the solution of the initial boundary value problem (2.1) in the case that the coefficient is chosen to be $\Lambda(s) = \alpha + \beta s$. If α and β are such that $a_0 \leq \alpha + \beta f(T) \leq a_1$, then Λ belongs to the class A and, by assumption (3.2), there is a T_0, $0 < T_0 \leq T$, such that

$$\partial_{xx} z(x,t) \geq 0 \quad \text{on } Q_{T_0}. \tag{3.6}$$

Let T_1 denote the smaller of the two numbers T_0 and t_1, and note that

$$\begin{aligned} \partial_t(u-z) - a(u)\partial_{xx}(u-z) &= (a(u) - \Lambda(u))\partial_{xx} z + (a'(u) - \Lambda'(u))(\partial_x z)^2 \\ &\quad + a'(u)(\partial_x u + \partial_x z)\partial_x(u-z) + \Lambda'(u)(u-z). \end{aligned}$$

That is, $w(x,t) = u(x,t) - z(x,t)$, satisfies

$$\begin{aligned}&\partial_t w - p(x,t)\partial_{xx}w - q(x,t)\partial_x w - r(x,t)w \\ &\quad = (a(u) - \Lambda(u))\partial_{xx}z + (a'(u) - \Lambda'(u))(\partial_x z)^2.\end{aligned}$$

Conditions (3.4) and (3.6) together imply that the right side of this equation is positive for $0 < x < 1$, $0 < t < T_1$. In addition,

$$\begin{aligned} w(x,0) &= 0, \qquad 0 < x < 1, \\ w(0,t) = w(1,t) &= 0, \qquad 0 < t < T_1, \end{aligned}$$

and thus the minimum principle implies that $w(x,t) > 0$ on Q_{T_1}. That is,

$$u(x,t) > z(x,t) \quad \text{for } 0 < x < 1, \quad 0 < t < T_1.$$

Similarly, we can show that

$$z(x,t) > v(x,t) \quad \text{for } 0 < x < 1, \quad 0 < t < T_1.$$

Hence,

$$\frac{u(x,t) - f(t)}{x - 0} > \frac{v(x,t) - f(t)}{x - 0} \quad \text{for } 0 < x < 1, \quad 0 < t < T_1.$$

Letting x tend to zero through positive values leads to the result

$$\partial_x u(0,t) > \partial_x v(0,t) \quad \text{for } 0 < t < T_1.$$

But this contradicts (3.5). Evidently the condition (3.3) is inconsistent with the existence of an interval in $[0, f(T)]$ on which a and b are separated by a linear graph. This proves that the data f and g uniquely determines the coefficient $a(u)$ within the equivalence class of polygonally equivalent coefficients. ∎

We are not yet able to prove the existence of an $a(u)$ in A given data $f \in D$ and g in $\mathcal{R}(N)$. However, we will show that if data f and g is given and if a coefficient $a(u)$ is assumed to exist such that $g = N[f;a]$, then a polygonal approximation for $a(u)$ can be constructed.

4 APPROXIMATING THE SOLUTION OF THE INVERSE PROBLEM

Let $f \in D$ be given and suppose that $g(t) = N[f;a]$ for some $a \in A$. Then, $a(0)$ is given in terms of f and g by (2.9). We let $A_0 = a(0)$. Fix a positive integer M and define a partition of the interval $[0,T]$:

$$0 = t_0 < t_1 < \cdots < t_M = T. \tag{4.1}$$

This induces a partition on $[0, f(T)]$:

$$0 = f(t_0) = f_0 < f_1 < \cdots < f_M = f(t_M) = f(T). \tag{4.2}$$

We seek to determine the numbers $A_k = a(f_k)$, $k = 1, 2, \dots, M$. We define a continuous, piecewise-linear approximation to $a(u)$ by

$$P_M[a](u) = \sum_{k=0}^{M} A_k \phi_k(u). \tag{4.3}$$

Here $\phi_0, \phi_1, \dots, \phi_M$ denote the well-known "hat" functions of approximation theory; i.e., each ϕ_k is linear on each interval $[f_{j-1}, f_j]$ with $\phi_k(f_j) = \delta_{jk}$, for $0 \le j, k \le M$. Note that Theorem 3.1 implies that, for any choice of the partition (4.1), then $N[f; a] = N[f; b]$ implies that $P_M[a] = P_M[b]$.

We introduce the following notation:

$$\begin{aligned} |a|_0 &= \sup\{|a(u)| : 0 \le u \le f(T)\}, \\ |a|_1 &= |a|_0 + |a'|_0, \\ |u|_T &= \sup_{Q_T} |u(x,t)|. \end{aligned}$$

Theorem 4.1 *Let $f \in D$ and $a \in A$ be given, and let $g = N[f; a]$. Also, for a fixed positive integer M and some partition (4.1), let $P_M[a]$ be determined by (4.3). Finally, let $u = u(x,t)$ and $U(x,t)$ denote the unique solutions of (2.1) corresponding to coefficients $a(u)$ and $P_M[a](u)$. Then,*

$$\begin{aligned} |u - U|_T &\le C_1 T |a - P_M[a]|_1, && (4.4a) \\ |\partial_x u - \partial_x U|_T &\le C_2 M \sqrt{T} |a - P_M[a]|_1, && (4.4b) \\ |\, N[f; a] - N[f; P_M(a)] \,|_0 &\le (C_1 T + a_1 C_2 M \sqrt{T}) |a - P_M[a]|_1. && (4.4c) \end{aligned}$$

Proof: It will suffice to prove the estimates on the set $Q' = [0,1] \times [0, t_1]$ where $P_M[u]$ is given by $A_0\phi_0(u) + A_1\phi_1(u) = A_0 + \beta_1 u$. The general estimates then follow by piecing together similar estimates on sets of the form $[0,1] \times [t_{j-1}, t_j]$.

Let $w(x,t) = u(x,t) - U(x,t)$, and note that it satisfies

$$\begin{aligned} &\partial_t w - a(u)\partial_{xx} w - a'(u)\partial_x(u+U)\partial_x w - \beta_1(\partial_{xx}U)w \\ &\qquad = (a(u) - P_M[u])\partial_{xx}U + (a'(u) - \beta_1)(\partial_x U)^2 \;\text{ in }\; Q', \\ &w(x,0) = 0, \quad 0 < x < 1, \\ &w(0,t) = w(1,t) = 0, \quad 0 < t < t_1. \end{aligned}$$

Then, $w(x,t)$ can be written in the form (Cannon [1])

$$\begin{aligned} w(x,t) &= \int_0^t \int_0^1 \Big[G(x,y,t-\tau)\Big((a(u) - P_M[u])\partial_{xx}U \\ &\qquad + (a'(u) - \beta_1)(\partial_x U)^2\Big)(y,\tau)\Big]\, dy\, d\tau. \end{aligned} \tag{4.5}$$

Note that

$$\beta_1 = \frac{a(f_1) - a(0)}{f_1 - 0} = a'(\xi) \;\text{ for some }\; \xi,\; 0 \le \xi \le f_1.$$

In the above, $G(x,y,t)$ denotes the appropriate fundamental solution for this linear parabolic initial boundary value problem. It follows from well-known estimates (Cannon [1]) for $G(x,y,t)$ that

$$|w|_{Q_{t_1}} \le C_1 t_1 |a - P_M[a]|_1.$$

Differentiating (4.5) with respect to x leads to the estimate

$$|w_x|_{Q_{t_1}} \le C_2\sqrt{t_1}|a - P_M[a]|_1.$$

By piecing together M estimates of this type we arrive at (4.4) and (4.4). Then (4.4), together with the relation $N[f;a] = -a(f(t))\partial_x u(0,t)$, leads to (4.4). ∎

We describe now an iteration scheme for generating a sequence of continuous, piecewise linear functions from given data f and g. If we assume that there exists a coefficient $a \in A$ such that $g = N[f;a]$, then we can show that the sequence converges monotonically to $P_M[a]$.

Note first that for given data $f \in D$ and $g \in G$, $A_0 = a(0)$ is determined by (2.7) and (2.8). We proceed next to generate a sequence of guesses $\{A_1^{(k)}\}$ that we will show to converge monotonically to $A_1 = a(f_1)$. Then, A_0 and A_1 determine $P_M[a](u) = A_0\phi_0(u) + A_1\phi_1(u)$ on the first subinterval $[f_0, f_1]$ of the partition (4.2). Continuing in this way, we construct, step by step, the M linear functions of which $P_N[a]$ is composed.

The algorithm for generating the iterates $\{A_j^{(k)}\}$, $j = 1,2,\ldots,M$, consists of the following steps:

For $j = 1,2,\ldots,M$:

1. For $m = 0,1,\ldots,j-1$, let $A_j^{(0)} = A_{j-1}$.
2. For $k = 0,1,\ldots$, let

   $$A_j^{(k)}(u) = A_{j-1}\phi_{j-1}(u) + A_j^{(k)}\phi_j(u) \text{ for } f_{j-1} \le u \le f_j.$$

   Denote the solution of the initial boundary problem (2.1) on $(0,1)\times(t_{j-1},t_j)$ corresponding to this linear coefficient by $U(x,t;\Lambda_j^{(k)})$.
3. Evaluate the next iterate:

   $$A_j^{(k+1)} = \frac{g(t_j)}{-\partial_x U(0,t_j;\Lambda_j^{(k)})}.$$

4. If $|\Lambda_j^{(k)} - \Lambda_j^{(k+1)}|$ is sufficiently small, then set $A_j = A_j^{(k+1)}$; otherwise return to step 2.

We can show that, for each j, the sequence $\{A_j^{(k)}\}$ is monotone.

Theorem 4.2 *Let $f \in D$ and $g \in \mathcal{R}(N)$ be given, and fix a positive integer M. Let $\{A_j^{(k)} : j = 1\ldots,M\}$ denote the M sequences of numbers generated by the algorithm described above. Then, for each j, $\{A_j^{(n)}\}$ is monotone as a function of the index n.*

Proof: Given data $f \in D$ and $g \in \mathcal{R}(N)$, $A_0 = a(0)$ is determined by (2.7). Then, according to the algorithm,

$$A_1^{(0)} = A_0 \quad \text{and} \quad \Lambda_1^{(0)}(u) = A_0\phi_0(u) + A_0\phi_1(u) = A_0.$$

From the solution $U(x,t;\Lambda_1^{(0)})$ of (2.1) on Q_{t_1} corresponding to the coefficient $a(u)$, we compute

$$A_1^{(1)} = -g(t_1)(\partial_x U(0,t_1;\Lambda_1^{(0)}))^{-1}. \tag{4.6}$$

Note that, if

$$-A_1^{(0)}\partial_x U(0,t_1;\Lambda_1^{(0)}) < g(t_1), \tag{4.7}$$

then (4.6) implies that

$$A_1^{(0)} < A_1^{(1)}. \tag{4.8}$$

If the inequality in (4.7) is reversed, then the direction of the inequality in (4.8) is likewise reversed. If (4.7) is an equality, then so is (4.8) and A_0 is a fixed point for the iteration.

Suppose for the sake of discussion that $A_1^{(1)} > A_1^{(0)}$ and let $U_1(x,t)$ and $U_0(x,t)$ denote the solutions of the initial boundary value problem (2.1) on Q_{t_1} corresponding to the coefficients

$$\begin{aligned} A_1^{(1)}(u) &= A_0\phi_0(u) + A_0^{(1)}\phi_1(u) = A_0 + \beta_1 u, \\ A_1^{(0)}(u) &= A_0\phi_0(u) + A_0^{(0)}\phi_1(u) = A_0. \end{aligned}$$

Let $w(x,t) = U_1(x,t) - U_0(x,t)$; then

$$\begin{aligned} &\partial_t w - (A_0 + \beta_1 U_1)\partial_{xx} w - \beta_1 \partial_x (U_1 + U_0)\partial_x w \\ &\quad = ((A_0 + \beta_1 U_1) - (A_0 + 0))\partial_{xx} U_0 + (\beta_1 - 0)(\partial_x U_0)^2 \\ &\quad = \beta_1(U_1 \partial_{xx} U_0 + (\partial_x U_0)^2), \\ &w(x,0) = 0, \quad 0 < x < 1, \\ &w(0,t) = w(1,t) = 0, \quad 0 < t < t_1. \end{aligned}$$

Since $\partial_{xx} U_0 > 0$ on Q_{t_1} it is evident that $\beta_1(U_1\partial_{xx}U_0 + (\partial_x U_0)^2)$ is positive on Q_{t_1} so that the minimum principle implies that $w(x,t) > 0$ on Q_{t_1}. Then

$$U_1(x,t) > U_0(x,t) \quad \text{on } Q_{t_1},$$

and it follows as in the proof of Theorem 3.1 that

$$0 < -\partial_x U_1(0,t_1) < -\partial_x U_0(0,t_1).$$

Thus, $A_1^{(1)} > A_1^{(0)}$ implies that $-\partial_x U_0(0,t_1) > -\partial_x U_1(0,t_1) > 0$, and

$$A_1^{(2)} = \frac{g(t_1)}{-\partial_x U_1(0,t_1} > \frac{g(t_1)}{-\partial_x U_0(0,t_1)} = A_1^{(1)}.$$

Thus, $A_1^{(1)} > A_1^{(0)}$ implies that $A_1^{(2)} > A_1^{(1)}$.

The same line of reasoning can be applied to complete an induction argument. Similarly, if $A_1^{(1)} < A_1^{(0)}$, the sequence of iterates decrease. ∎

If we suppose that there exists an $a \in A$ such that $g = N[f; a]$ for the given data f and g, then we can show that each of the monotone sequences $\{A_j^{(n)}\}$ is convergent to $A_j = a(f_j)$. Thus, if a is assumed to exist, the algorithm generates $P_M[a]$.

Theorem 4.3 *Let $f \in D$ and $g \in \mathcal{R}(N)$ be given and suppose that $g = N[f; a]$ for some $a = a(u)$ in A. For a fixed M and a fixed partition $\{f_0, f_1, \ldots, f_M\}$ of $[0, f(T)]$, let $\{A_j^{(n)}\}$, $j = 1, \ldots, M$, denote the sequences generated by the algorithm. Then for each j, $\{A_j^{(n)}\}$ converges monotonically to $A_j = a(f_j)$.*

Proof: Suppose for the sake of discussion that $A_1 > A_0$ and let $U_1(x,t)$ and $U_0(x,t)$ denote the solutions of the initial boundary value problem (2.1) corresponding to coefficients $A_0\phi_0(u) + A_1\phi_1(u) = A_0 + \beta_1 u$ and $A_0\phi_0(u) + A_1^{(0)}\phi_1(u) = A_0$, respectively. Since $A_1 > A_0$, it follows that

$$\beta_1 = \frac{A_1 - A_0}{f_1 - f_0} > 0 \text{ and } A_0 + \beta_1 u > A_0 \text{ for } f_0 < u < f_1.$$

Then we carry out a comparison argument based on the maximum principle to show that $A_1^{(1)} < A_1$. Similarly, we can show $A_1^{(n)} < A_1$ for $n = 0, 1, \ldots$. Then A_1 is an upper bound for the monotone increasing sequence $\{A_1^{(n)}\}$.

If A_1 is not the *least* upper bound for the sequence, then the sequence converges to a limit $A' < A_1$. But in this case the coefficients $A_0\phi_0(u) + A_1\phi_1(u)$ and $A_0\phi_0(u) + A_1'\phi_1(u)$ generate solutions $u(x,t)$ and $u'(x,t)$, respectively, for the initial boundary value problem (2.1). Since the sequence converges to A', the solutions satisfy the condition

$$-(A_0\phi_0(f_1) + A_1\phi_1(f_1))\partial_x u(0,t_1) = -(A_0\phi_0(f_1) + A_1'\phi_1(f_1))\partial_x u'(0,t_1),$$

which implies that $\partial_x u'(0,t_1) > \partial_x u(0,t_1)$. But, if

$$A_0\phi_0(u) + A_1\phi_1(u) > A_0\phi_0(u) + A_1'\phi_1(u) \text{ for } f_0 < u < f_1,$$

the comparison argument implies that $u'(x,t_1) < u(x,t_1)$ for $0 < x < 1$. This leads to the same contradiction encountered in the proof of Theorem 3.1 and we conclude that A_1 is the least upper bound for the sequence.

This proves that $\{A_1^{(n)}\}$ converges monotonically to A_1. In a similar way, we can show that $\{A_j^{(n)}\}$ converges monotonically to $A_j = a(f_j)$ for each j, $j = 1, \ldots, M$. This completes the proof in the case that the slopes

$$\beta_j = \frac{A_j - A_{j-1}}{f_j - f_{j-1}}, \; j = 1, 2, \ldots, M,$$

are all positive. The proof in the more general case where the β_j are not all positive uses the same arguments with the difference that, when β_j is negative, the sequence $\{A_j^{(n)}\}$ decreases monotonically to A_j instead of increasing. ∎

5 NUMERICAL EXPERIMENTS

We will describe the results of some numerical experiments in which the algorithm of the previous section is applied to the inverse problem (3.1). For

$$f(t) = Ct^2, \quad C > 0,$$

and the functions

$$a(u) = A_0\big(2 + 2/\pi \arctan(K(u+\mu))\big), \tag{5.1}$$

$$a(u) = \frac{A_0}{1 + K^{2p}(u+\mu)^{2p}}, \tag{5.2}$$

$$a(u) = A_0 \exp(-K^{2p}(u+\mu)^{2p}), \tag{5.3}$$

the initial boundary value problem (2.1) was solved numerically for the unknown function $u = u(x,t)$. Then the numerical solution $u(x,t)$ was used to compute the data $g(t)$ from (2.2).

Before submitting the data f and g to the algorithm for the recovery of the coefficient $a = a(u)$, random error was superimposed on the data functions to obtain "pseudo data" f^* and g^* given by

$$f^*(t) = f(t)(1 + e_1(t)) \text{ and } g^*(t) = g(t)(1 + e_2(t)),$$

where e_1 and e_2 denote zero mean, normally distributed random variables such that $|e_k| \le E_k$, $k = 1,2$. The relative error levels E_k were varied from zero up to .10, simulating relative error of zero to ten percent.

Figure 1 shows the graph of the coefficient $a(u)$ given by (5.1) superimposed on the graph of the coefficient recovered from the algorithm. The parameter values for (4.2) were $A_0 = .5$, $K = 16$, and $\mu = -.1$, and the error levels in f and g were 5 and 10 percent, respectively.

Figures 2 and 3 provide similar plots for $a(u)$ given by (5.2) and (5.3) with the parameter values given by:

$$A_0 = 1, \quad K = 48, \quad \mu = -.07, \text{ and } p = 1 \text{ in (5.2)}$$
$$A_0 = 1, \quad K = 12, \quad \mu = -.07, \text{ and } p = 3 \text{ in (5.3)}$$

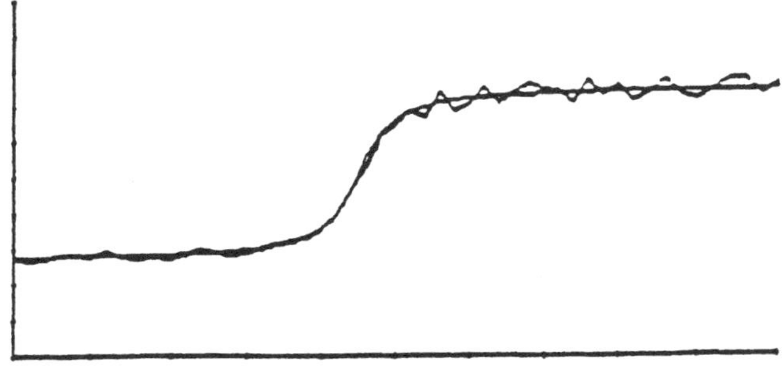

Figure 1:

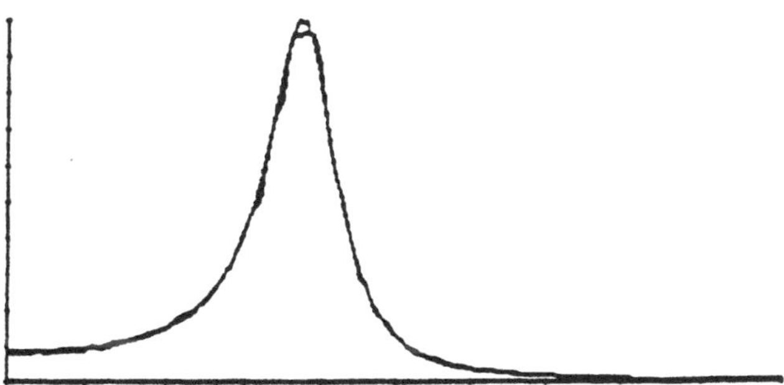

Figure 2:

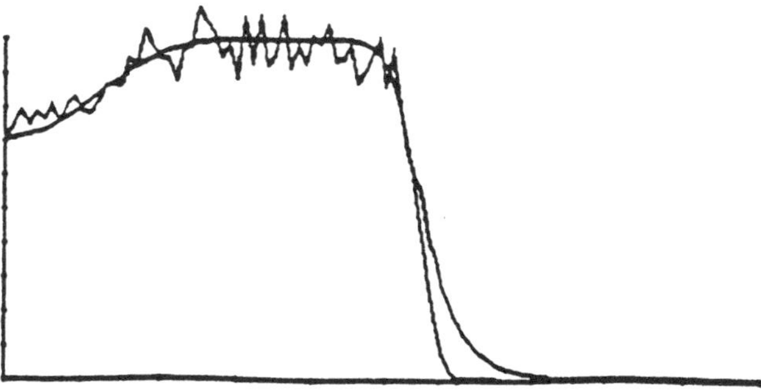

Figure 3:

REFERENCES

[1] Cannon J. R. *The One Dimensional Heat Equation.* Addison Wesley Pub. Corp., 1984.

[2] Cannon J. R., DuChateau P. Determination of unknown coefficients in parabolic operators from overspecified initial and boundary data. *Jour. of Heat Transfer*, 100:503–507, 1978.

[3] Cannon J. R., DuChateau P. Some asymptotic boundary behavior of solutions of nonlinear parabolic initial boundary value problems. *JMAA*, 68(2):536–547, 1979.

[4] Cannon J. R., DuChateau P., Steube K. Trace type functional differential equations and the identification of hydraulic properties of porous media. *Transport in Porous Media*, 6:745–758, 1991.

[5] DuChateau P. Boundary behavior and monotonicity estimates for solutions to nonlinear diffusion equations. *Rocky Mtn. Jour.*, 15(4):769–786, 1985.

[6] Gilding B.H. Improved theory for a nonlinear degenerate parabolic equation. *Ann. Scuola Norm. Sup. Pisa*, 16(4), 1989.

International Series of Numerical Mathematics, Vol. 114, © 1993 Birkhäuser Verlag Basel

REACTIVE FLOWS IN POROUS MEDIA: THE REACTION-INFILTRATION INSTABILITY

John Chadam*

Abstract. In this note we summarize some recent results of reactive flows in porous media for which the fluid/solid reaction can increase the porosity/permeability of the medium. The phenomenon is modelled by a coupled system of nonlinear ordinary and partial differential equations for which a global existence and uniqueness theorem is stated. As a physically relevant parameter tends to zero the problem converges to a moving free boundary problem. The shape stability of planar reaction interfaces is discussed in this context using local bifurcation analysis. Numerical results are presented to show the complexity of the possible evolving reaction interfaces.

1 INTRODUCTION

When reactive fluids flow through a porous medium they can dissolve some of the minerals and cause changes in the porosity and permeability. Through Darcy's law the resulting alterations in the flow pattern can give rise to a *reaction-infiltration instability* [1, 2, 4] which can cause instabilities in the shape of the porosity level curves (e.g., fingering). The crucial destabilizing feedback mechanism works as follows. If a protrusion (in the porosity level curves) in the reaction zone exists at some time, the flow of the most reactive fluid tends to be focussed at the tip of the protrusion via Darcy's law since "inside" the protrusion (the upstream side) the permeability is greater than in the neighboring regions (see Figure 1). Since additional reactive fluid now arrives at the tip, it tends to advance more rapidly, causing fingering. On the other hand, diffusion from the sides of the protrusion causes the fluid which is focussing to the tip to be less reactive and hence to decelerate its advancement. The competition between these two mechanisms results in shape selection (self-organization) of the reaction zone leading either to the decay of the protrusion, restabilization to a morphologically more complicated advancing reaction zone (fingering) or the temporal development of successively more complicated patterns (tip splitting, budding, meandering, etc.). Understanding this shape selection process has important applications to many geochemical situations (e.g., the diagenesis and evolution of mineral deposits, oil and gas reservoirs, the dynamics of breakout from chemical and nuclear waste repositories, in situ coal gasification, enhanced oil recovery, the location of roll-front deposits, the leaching of minerals, etc. [1, 2]).

In this note we shall give a summary of the main results of the author and many collaborators obtained in the study of the mathematical model of this phenomenon and state some open problems. The mathematical model (in the form of coupled sets of ordinary and partial differential equations) is given in §2 along with an announcement of recent rigorous existence, uniqueness and asymptotic results. In §3 we study the

*The Fields Institute, 185 Columbia St., Waterloo, Ontario, N2L 5Z5, Canada

Figure 1: Focusing of Flow to the Tip of a Porosity Level Curve.

convergence of this model to a moving free-boundary problem as the geologically relevant parameter (the ratio of the solute concentration to dissolved solute density) tends to zero. The next section, §4, is devoted to the above-mentioned problem of the shape stability of the reaction interface using bifurcation analysis and numerical simulation. In §5, a summary of these results is given and mention is made of their extension to a) the situation when viscosity changes can also occur during the reaction (leading to a coupling of the present reaction-infiltration instability with the Saffman-Taylor instability [7]) and b) the case when the porous medium is layered.

2 MATHEMATICAL MODEL AND THEORETICAL RESULTS

The domain of the model is assumed to be the infinite strip: $(x, y) \in (-\infty, \infty) \times [0, \pi]$. Within this region the coupled set of nonlinear equations which models these processes is (see [4] for more details):

$$\varphi_t = -k(\varphi_f - \varphi)^{2/3}(c - c_{eq}), \tag{2.1a}$$

$$(\varphi c)_t = \nabla \cdot [\varphi D(\varphi) \nabla c + \varphi c \kappa(\varphi) \nabla p] + \rho \varphi_t, \tag{2.1b}$$

$$\varphi_t = \nabla \cdot [\varphi \kappa(\varphi) \nabla p], \tag{2.1c}$$

where the unknowns are φ, c, p, the porosity, concentration of solute in water and pressure, respectively. Here, the reaction rate coefficient, κ, the equilibrium concentration of solute in water, c_{eq}, and the density of dissolving mineral, ρ, are constants, while D and κ are porosity-dependent diffusion and permeability coefficients. The 2/3 power in (2.1*a*) indicates that we are considering surface reactions. Darcy's law for the velocity $\vec{v}$ has been used in (2.1*b*,*c*) in the form (the viscosity is taken to be constant) $\vec{v} = -\kappa(\varphi) \nabla p$. Equations (2.1) are to be solved for φ, c, p subject to the imposed initial data and the following asymptotic conditions:

$$c \to 0,\ \varphi \to \varphi_f \text{ and } p_x \to p_f' \text{ as } x \to -\infty, \tag{2.2a}$$

$$c \to c_{eq},\ \varphi \to \varphi_0 \text{ and } p_x \text{ to be determined as } x \to +\infty. \tag{2.2b}$$

In typical geological situations, $\varepsilon = c_{eq}/\rho$ is quite small (e.g. from 10^{-3} to 10^{-10}). As we shall see, this will force the zone where all the complicated, and from our present viewpoint, unimportant reactions occur to be very thin (indeed $0(\varepsilon^{1/2})$) and to move very slowly. To capture this behavior we transform equation (2.1) using

$$\begin{array}{ll} \tilde{t} = \varepsilon(kc_{eq})t, & \tilde{\underline{r}} = [kc_{eq}/D(\varphi_f)]^{1/2}\underline{r}, \\ \tilde{c} = c/c_{eq}, & \tilde{p} = \kappa(\varphi_f)p/D(\varphi_f), \\ \tilde{D}(\varphi) = D(\varphi)/D(\varphi_f), & \tilde{\kappa}(\varphi) = \kappa(\varphi)/\kappa(\varphi_f) \end{array}$$

to obtain (dropping the tildes)

$$\varepsilon(\varphi c)_t = \nabla \cdot (\varphi D(\varphi) \nabla c + \varphi\kappa(\varphi)c \nabla p) + \varphi_t, \tag{2.3a}$$

$$\varepsilon\varphi_t = -(\varphi_f - \varphi)^{2/3}(c-1), \tag{2.3b}$$

$$\varepsilon\varphi_t = \nabla \cdot (\varphi\kappa(\varphi) \nabla p), \tag{2.3c}$$

with the corresponding scaled versions of equation (2.2).

We would like to announce the following existence and uniqueness results for the coupled system of nonlinear parabolic, elliptic and ordinary differential equations (2.3) with homogeneous Neumann conditions on $y = 0, \pi$.

Theorem 2.1 *(X. Chen and J. Chadam) Suppose* $\varepsilon > 0$ *and*

$$(x, y) \in [-M, M] \times [0, \pi] = \Omega_M.$$

There exists a global (in time) weak solution to equations (2.3) with

$$\begin{array}{c} c \in L^\infty((0,T) \times \Omega_M) \cap L^2(0,T; H^1(\Omega_M)), \\ \varphi \in C^{1/2}(0,T; L^\infty(\Omega_M)), \\ p \in L^\infty(0,T; H^1(\Omega_M)) \quad \text{for any} \quad T < \infty, \end{array}$$

which is unique up to an additive constant in p.

In addition, a complete regularity theory has also been established for these equations. Similar results with $(c, \varphi, p) \in L^2_{loc}((0,T); L^2(\Omega_M))$ were announced for bounded rectangles but with Dirichlet data by Collet [6]. For an aquifer which is infinite in the x-direction, one expects global existence but, to date, with X. Chen we have only been able to establish this up to time T_0/ε where $T_0 < \infty$ is estimated from the data. In the next section, we shall describe our results for the physically relevant limit (large solid density asymptotics) as $\varepsilon \to 0$.

3 THE MOVING FREE BOUNDARY MODEL

Since, for geological reasons, we are primarily interested in shape instabilities of the thin reaction zone, it is pragmatic to collapse this region which contains the complicated chemistry into a moving front by taking $\varepsilon \to 0$. The effect of the internal chemistry will manifest itself as jump conditions across this moving reaction interface. The jump conditions will in turn govern the evolution of the interface. Formally [4], one uses multiple scale analysis and expands the unknowns c, φ, p (f stands for any of these) in outer and inner expansions of the form

$$f^{outer} = f_0(x, y, t) + \varepsilon^{1/2} f_1(x, y, t) + \cdots,$$

$$f^{inner} = \tilde{f}_0(\sigma/\varepsilon^{1/2}, \tau, t) + \varepsilon^{1/2}\tilde{f}_1(\sigma/\varepsilon^{1/2}, \tau, t) + \cdots,$$

where σ and τ are coordinates normal and tangential to the thin $(O(\varepsilon^{1/2}))$ moving reaction zone. Using standard matching conditions, one obtains at the lowest order, with the location of the moving free-boundary (reaction front) being $x = R(y,t)$:

In the upstream region, $x < R(y,t)$,

$$\varphi = \varphi_f, \tag{3.1a}$$

$$\Delta c + \nabla c \cdot \nabla p = 0, \tag{3.1b}$$

$$\Delta p = 0; \tag{3.1c}$$

In the downstream region, $x > R(y,t)$,

$$\varphi = \varphi_0, \tag{3.1d}$$

$$c = 1, \tag{3.1e}$$

$$\Delta p = 0; \tag{3.1f}$$

And on the unknown reaction front, $x = R(y,t)$,

$$c = 1, \tag{3.1g}$$

$$p_- = p_+, \tag{3.1h}$$

$$\left(\frac{\partial p}{\partial x} - \frac{\partial p}{\partial y}\frac{\partial R}{\partial y}\right)_- = \Gamma\left(\frac{\partial p}{\partial x} - \frac{\partial p}{\partial y}\frac{\partial R}{\partial y}\right)_+, \tag{3.1i}$$

$$\frac{\partial c}{\partial x} - \frac{\partial c}{\partial y}\frac{\partial R}{\partial y} = (1 - \varphi_0/\varphi_f)R_t, \tag{3.1j}$$

where the subscripts $\pm$ indicate evaluation on $x = R \pm 0$, and $\Gamma = \varphi_0\kappa_0/\varphi_f\kappa_f$ is a constant in (0,1] which measures the amount of dissolution caused by the completed reaction. Equation (3.1*j*) is the Stefan condition giving, after division of both sides by $(1 + R_y^2)^{1/2}$, the normal velocity of the front in terms of the normal flux of the concentration. These equations are to be solved along with the asymptotic conditions

$$c \to 0,\ \varphi \to \varphi_f \text{ (trivially) and } p_x \to p_f' \text{ as } x \to -\infty, \tag{3.2a}$$

$$c \to 1,\ \varphi \to \varphi_0 \text{ (both trivially) and } p_x \text{ to be determined as } x \to +\infty. \tag{3.2b}$$

Before using the moving free-boundary model (3.1) to solve the shape stability problem for the interface $x = R(y,t)$, we would first like to announce some preliminary results on the rigorous justification of the limiting procedure $\varepsilon \to 0$ to obtain equations (3.1). All of our results to date are for one dimension; the two-dimensional versions are open problems currently under investigation.

Theorem 3.1 *(X. Chen, J. Chadam, R. Gianni, R. Ricci) For $\varepsilon > 0$ and $x \in (-\infty,\infty)$, the one-dimensional version of equations (2.3) have a global weak solution which is unique up to an additive constant in p.*

Fundamentally, the reason for being able to prove global existence in one dimension, but not yet in two, is that equation (2.3c) is immediately integrable in terms of φ and c, giving a p-reduced version of equations (2.3).

Theorem 3.2 *(X. Chen, J. Chadam, R. Gianni, R. Ricci) Suppose $(c_\varepsilon, \varphi_\varepsilon)$ is a global solution of the reduced version of equations (2.3). There exists a subsequence $\{\varepsilon_j\}$ such that, as $\varepsilon_j \to 0$, $c_{\varepsilon_j} \rightharpoonup c$ and $\varphi_{\varepsilon_j} \rightharpoonup \varphi$ weakly in $L^2([-M,M] \times (0,T))$ for all M, T,* *where c and φ are solutions of the p-reduced, one-dimensional versions of equations (3.1) (i.e., having solved (3.1f,h,i) along with $p_x \to p'_f$ as $x \to -\infty$ explicitly for p and substituted this solution into (3.1b)).*

When φ_0 is constant (the case of an initially homogeneous porous medium which is of central importance for showing the autonomous (no external pattern) generation of patterned fronts), the solution of the one-dimensional version of (3.1) is the travelling wave:

$$x = \overline{R}(t) = V(t), \tag{3.3a}$$

$$\bar{c}(x,t) = \begin{cases} e^{-V(x-Vt)}, & x < Vt, \\ 1, & x > Vt, \end{cases} \tag{3.3b}$$

$$\bar{p}(x,t) = \begin{cases} p'_f(x-Vt), & x < Vt, \\ p'_f(x-Vt)/\Gamma, & x > Vt, \end{cases} \tag{3.3c}$$

where the velocity V of the travelling wave is given by $V = v_f(1-\varphi_0/\varphi_f)^{-1}$ for an inlet velocity v_f. It is instructive to examine the convergence described in Theorem 3 in the special case of travelling wave solutions of the p-reduced versions of equations (2.3). With $\xi = x - V_\varepsilon t$ and prime denoting differentiation with respect to ξ, after some manipulation the ordinary equations can be written as

$$\varphi' = \frac{1}{\varepsilon V_\varepsilon}(\varphi_f - \varphi)^{2/3}(c-1), \tag{3.4a}$$

$$c' = \frac{V_\varepsilon}{\varphi D(\varphi)}[c(\varphi_f - \varphi_0) - (\varphi_f - \varphi)], \tag{3.4b}$$

and one would like to prove the existence of an orbit connecting $(\varphi, c) = (\varphi_f, 0)$ to $(\varphi_0, 1)$. By analogy with $y' = y^{2/3}$, the solution for $\xi < 0$ is

$$\varphi = \varphi_f, \qquad c = \exp\left[\frac{V_\varepsilon(\varphi_f - \varphi_0)\xi}{\varphi_f D(\varphi_f)}\right]\gamma_\varepsilon, \tag{3.5}$$

where γ_ε is the parameter which must be chosen so that the equation (3.4) can then be solved in $\xi > 0$ with $(\varphi(0), c(0)) = (\varphi_f, \gamma_\varepsilon)$ in order to reach $(\varphi_0, 1)$ at $\xi = +\infty$. A standard phase plane argument then gives the following result.

Theorem 3.3 *(X. Chen, E. Comparini, R. Ricci, J. Chadam) For $\varepsilon > 0$ there exists a unique γ_ε and hence a unique solution $(\varphi_\varepsilon, c_\varepsilon, V_\varepsilon)$ of problem (3.4). Moreover, there exist constants c_1, $c_2 > 0$ such that $c_1\sqrt{\varepsilon} < 1 - \gamma_\varepsilon < c_2\sqrt{\varepsilon}$. Finally, if (φ, c) is the solution (3.3) of the p-reduced equations (3.1), then $c_\varepsilon \to c$ in $C_0(-\infty, \infty)$ of order $\sqrt{\varepsilon}$, $\varphi_\varepsilon \equiv \varphi$ in $\xi < 0$ and $\varphi_\varepsilon \to \varphi$ of order $\sqrt{\varepsilon}$ in $L^1(0,\infty)$, the convergence being uniform in any interval $[\delta, \infty)$, $\delta > 0$, and finally $V_\varepsilon \to V$ of order ε.*

These results give an indication why $\sqrt{\varepsilon}$ was used in the original inner and outer expansions. Moreover, it should also be noted that it is a non-constructive existence result for equations (2.3). On the other hand one can compute the explicit solution (3.3) of the $\varepsilon \to 0$ limit of the planar versions of (2.3) and, using perturbation theory, study the shape stability of planar fronts. This will be done in the context of (3.1) and (3.3) in the next section.

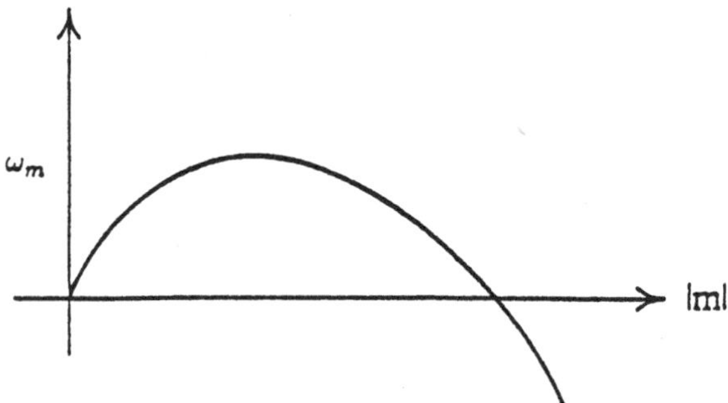

Figure 2: A Graph of ω_m vs. $|m|$.

4 SHAPE STABILITY OF PLANAR FRONTS

In doing a linearized stability analysis of the planar solutions (3.1), one considers solutions of (3.3) of the form

$$\begin{aligned} R(y,t) &= \overline{R}(t) + \delta R(y,t), \\ c(x,y,t) &= \bar{c}(x,t) + \delta c(x,y,t), \\ p(x,y,t) &= \bar{p}(x,t) + \delta p(x,y,t), \end{aligned}$$

where δR, δc and δp are small, leading to linearized versions of (3.3) for these unknowns. Since these equations are linear, one may decompose into the Fourier modes $\{\cos my\}$ and take Laplace transforms in t. That is, by expressing the perturbations as

$$\begin{aligned} \delta R(y,t) &= e^{\omega_m t}\cos my, \\ \delta c(x,y,t) &= C_m(x)e^{\omega_m t}\cos my, \\ \delta p(x,y,t) &= P_m(x)e^{\omega_m t}\cos my, \end{aligned}$$

we obtain a system of decoupled, second-order, linear, constant-coefficient ode's for $C_m(x)$ and $P_m(x)$. These can be solved directly and the Stefan condition (3.1j) gives an explicit expression for ω_m, the spectrum of the linearized problem [4]:

$$\omega_m = \frac{\lambda}{(1+\Gamma)(1-\varphi_0/\varphi_f)}[\lambda - (\lambda + 4m^2)^{1/2} + (1-\Gamma)|m|],$$

where $\lambda = v_f/D_f$ is the Peclet number. The graph of ω_m given in Figure 2 indicates that long wavelength perturbations are unstable, while short wavelengths are stabilized by the diffusion. The critical wave number $|m_0|$, where $\omega_m = 0$, is given by

$$|m_0| = \frac{2(1-\Gamma)}{(3-\Gamma)(1+\Gamma)}\,\lambda. \tag{4.1}$$

Since the channel of width π can only accommodate the modes $\{\cos my\}$, $m = 1,2,3,\ldots$, we obtain a sequence of critical values for the bifurcation parameter λ,

$$\lambda_{c,m} = \frac{(3-\Gamma)(1+\Gamma)}{2(1-\Gamma)}, \qquad m = 1,2,3,\ldots, \tag{4.2}$$

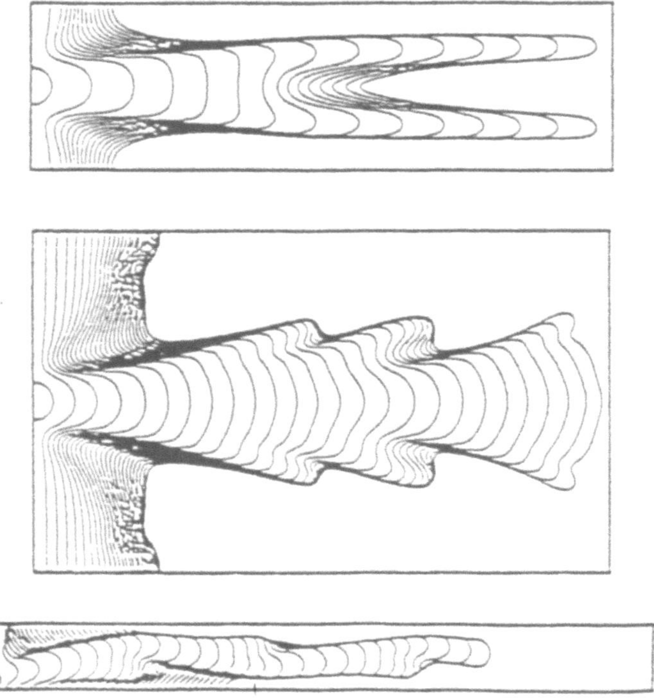

Figure 3: Numerical Simulations Exhibiting a) Tip-Splitting, b) Budding and c) Meandering (from Chen and Ortoleva).

at which successively higher modes become unstable relative to the planar state. It should be noted from (4.1) and (4.2) that when $\Gamma = 1$ (i.e., there is no porosity change) the planar state remains stable to all shape perturbations.

Of course, the above analysis only indicates the onset of shape instabilities from the planar state and says nothing about the global (fully nonlinear) bifurcation diagram containing secondary and higher order bifurcations. This is a difficult problem for equations (3.1), but a good indication of the wide variety of ensuing static and dynamic patterns that are possible can be seen (Figure 3) from the following numerical simulations of equations (2.3) with ε small done by Chen and Ortoleva [5].

5 CONCLUSIONS

We have given a mathematical model (2.3) describing the flow of reactive fluids in a porous medium which allows for the possibility of porosity changes due to the reaction with the medium. The basic existence and uniqueness results for these equations were summarized, as well as a statement of how this model converges to a moving free-boundary problem as $\varepsilon = c_{eq}/p \to 0$. The shape stability of planar fronts is examined in the context of the $\varepsilon = 0$, free-boundary model using bifurcation theory. One finds analytically that the reaction-infiltration feedback mechanism does indeed destabilize planar fronts to long wavelength perturbations. However, it is essential that a porosity change occur for this mechanism to become operative. Finally, it should be noted that the above analysis can be extended to reactions which cause viscosity changes as well (i.e., a coupling of the reaction-infiltration instability with the Saffman-Taylor instability), and, as expected, the planar reaction fronts are shown to be less stable [3]. Furthermore, the above models can be extended, using homogenization methods, to layered media and interesting results emerge on whether the instability is enhanced or repressed depending on the type of layering [8]. It was interesting to learn at the Oberwolfach workshop that the present methods might also apply to models in which the change in diffusion and permeability are due to the varying water content. A similar destabilizing feedback mechanism can be described in this situation.

Acknowledgement. Research supported by NSERC (Canada).

REFERENCES

[1] Chadam J., Hettmer J., Merino E., Moore C., Ortoleva P. Geochemical self-organization I: Feedback mechanisms and modelling approach. *Amer. J. Sci.*, 287:977–1007, 1987.

[2] Chadam J., Merino E., Ortoleva P., Sen A. Self-organization in water-rock interaction systems II: The reaction-infiltrate instability. *Amer. J. Sci.*, 287:1008–1040, 1987.

[3] Chadam J., Ortoleva P., Peirce A. Stability of reactive flows in porous media: Coupled porosity and viscosity changes. *SIAM J. Appl. Math.*, 51:684–692, 1991.

[4] Chadam J., Ortoleva P., Sen A. Reactive percolation instability. *IMA J. Appl. Math.*, 36:207–220, 1987.

[5] Chen W., Ortoleva P. Reaction front fingering in carbonate-cemented sandstones. *Earth-Sci. Rev.*, 29:183–198, 1990.

[6] Collet J.-F. *Construction of weak solutions in a two-dimensional domain for a problem arising in hydrogeology.* PhD thesis, Indiana University, 1992.

[7] Saffman P. G., Taylor G.I. The penetration of a fluid into a porous medium or Hele-Shaw cell containing a more viscous liquid. *Proc. Roy. Soc. London Ser. A*, 245:312–329, 1958.

[8] Xin J., Peirce A., Chadam J., Ortoleva P. Reactive flows in layered porous media II. The shape instability of the reaction interface. *SIAM J. Appl. Math.* To appear.

International Series of Numerical Mathematics, Vol. 114, © 1993 Birkhäuser Verlag Basel

A Study of the Effect of Inhomogeneities on Immiscible Flow in Naturally Fractured Reservoirs

Jim Douglas, Jr.* Jeffrey L. Hensley† Paulo Jorge Paes Leme‡

Abstract. The so-called medium block model for two-phase, immiscible, incompressible flow in a naturally-fractured petroleum reservoir is extended to admit inhomogeneous and non-periodic physical properties in both fractures and matrix blocks. In particular, a number of block types is allowed over each point in the reservoir, along with inhomogeneous physical properties. Numerical studies are performed to analyze the dependence of the flow on these properties for both vertical cross-sections and five-spot injection.

1 INTRODUCTION

Flow in naturally fractured petroleum reservoirs has been simulated for several decades by means of a variety of *double porosity* models. These models were first put forward on an *ad hoc* basis ([4], [6], [19]); more recently, models have been derived through the mathematical technique of homogenization (or other essentially equivalent averaging procedures); see [2], [7], [8]. Most of these recent models extend some simpler, single phase models discussed by Arbogast [1] and Showalter [17]. Our object is to consider certain generalizations of one such model for immiscible, incompressible, two-phase flow ([2], [7], [9]) in order to allow the introduction of several types of inhomogeneities into the description of the physical medium. In particular, it is not always realistic to assume that the matrix blocks in the fractured medium in the neighborhood of a point in the reservoir are all alike, and it is rarely realistic to assume that any overall periodicity exists in physical properties in either the fractures or the matrix blocks. More specifically, we shall develop a model admitting a number of block types in the neighborhood of any point in the reservoir, as well as inhomogeneous porosity, permeability, relative permeability, and capillary pressure functions in the fractures and, to some extent, in the blocks.

We shall begin by summarizing the concepts of the derivation from micromodel to macromodel of what the current authors and some of our colleagues have called a "medium block model" for immiscible flow in fractured media. Then, we shall introduce the generalizations we wish to consider and discuss the physical bases for these modifications in the underlying model. Lastly, we shall present the results of a collection of numerical studies of the effects of the inhomogeneities admissible under the extended model. Both vertical cross-section and "five-spot" examples will be treated.

*Department of Mathematics, Purdue University, West Lafayette, IN 47907-1395, USA

†Center for Parallel and Scientific Computing, University of Tulsa, Tulsa, OK 74104-3189

‡Instituto Politécnico, Universidade do Estado do Rio de Janeiro, 28600 Nova Friburgo - RJ and Departamento de Matemática, Pontifícia Universidade Católica do R.J., 22453 Rio de Janeiro - RJ, Brazil

2 A MEDIUM BLOCK MODEL

The medium block model considered in [2], [7], and [9] is based on a periodic structure formed by three families of parallel, equally spaced, but not necessarily orthogonal, fracture planes. Specifically, let the reservoir $\Omega \subset \mathcal{R}^3$ be a connected domain with a periodic structure, where the standard period is a cell (parallelepiped) $\mathcal{Q}$ consisting of a matrix block domain $\mathcal{Q}_m$ completely surrounded by a connected fracture domain $\mathcal{Q}_f$ (see Figure 1).

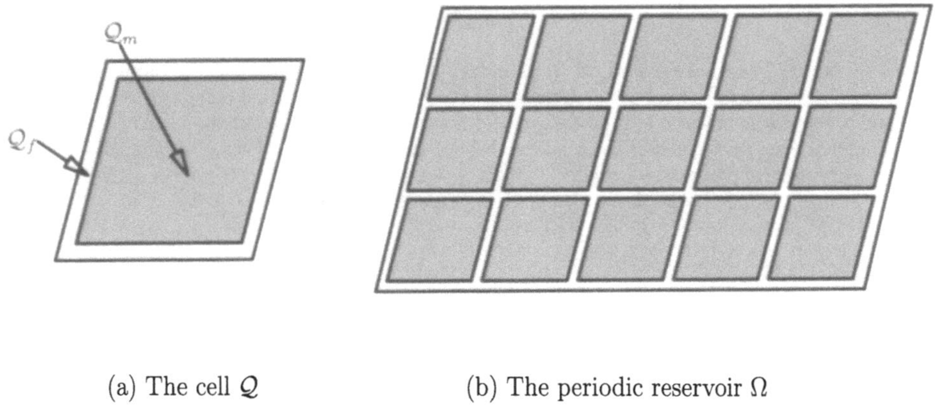

(a) The cell $\mathcal{Q}$ (b) The periodic reservoir Ω

Figure 1: The standard periodic cell and the periodic reservoir Ω.

We shall scale this periodic structure by a parameter $\epsilon > 0$; $\epsilon = 1$ corresponds to the original reservoir as described above. The ϵ-reservoir consists of copies of $\epsilon\mathcal{Q}$ covering Ω. Each ϵ-cell is adjacent to, but not overlapping, its neighbors (see Figure 2). There are four (three scaled) regions of interest: the fixed external boundary of the reservoir $\partial\Omega$, the set Ω_f^ϵ of fractures, the collection Ω_m^ϵ of matrix blocks, and the matrix-fracture interface Γ^ϵ, where

$$\Omega_f^\epsilon = \Omega \cap \bigcup_{\xi \in \mathcal{A}} \epsilon(\mathcal{Q}_f \cup \partial\mathcal{Q} + \xi),$$

$$\Omega_m^\epsilon = \Omega \cap \bigcup_{\xi \in \mathcal{A}} \epsilon(\mathcal{Q}_m + \xi),$$

$$\Gamma^\epsilon = \Omega \cap \bigcup_{\xi \in \mathcal{A}} \epsilon(\partial\mathcal{Q}_m + \xi),$$

and $\mathcal{A}$ is an appropriate infinite lattice.

Denote the oil, or nonwetting, phase by the index o and the water, or wetting phase, by w; let their densities and viscosities be ρ_α and μ_α, $\alpha = o, w$, respectively. Then, the equations (see, e.g., [3], [5], [16]) that govern saturated, two-phase, incompressible, immiscible flow in a single porosity system can be put into a convenient form as below. Let $s(x,t)$ denote the w-saturation, so that the o-phase has saturation $1 - s$. Let $p_\alpha(x,t)$, $\alpha = o, w$, represent the pressure in the α-phase, and denote the capillary

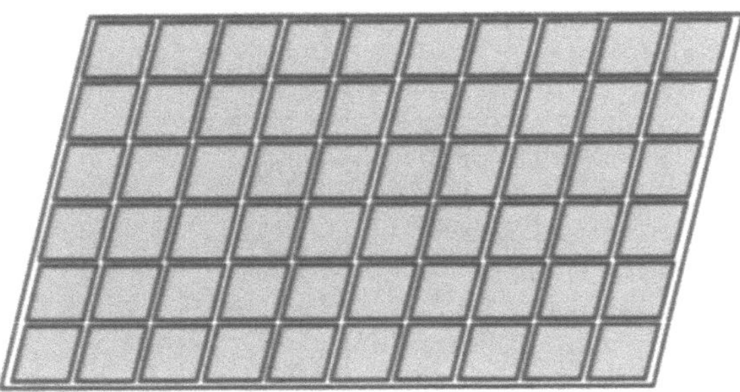

Figure 2: The ϵ-reservoir ($\epsilon = \frac{1}{2}$).

pressure between the two phases by

$$p_c(s) = p_o - p_w;$$

$p_c(s)$ is assumed to be a function of s only and typically is a decreasing function of s. It becomes infinite as the saturation tends to the residual water saturation $s_{\min} = s_{rw}$, and it is zero at the residual oil saturation corresponding to $s = s_{\max} = 1 - s_{ro}$.

Relative permeability functions, $k_{r\alpha}(s)$, $\alpha = o, w$, quantify the interference to flow in each phase caused by the presence of the other. Usually, $k_{rw}(s_{\min}) = k_{ro}(s_{\max}) = 0$. If $\mathbf{k}(x)$ is the absolute permeability tensor, then $\mathbf{k}k_{r\alpha}$ is the permeability of the rock to the α-phase at the point x with saturation s.

Potentials are more convenient to work with than pressures; denote them by

$$\psi_\alpha = p_\alpha - \rho_\alpha g z, \quad \alpha = o, w,$$

where g is the gravitational constant (not vector) and $z(x)$ is the depth. Muskat's extension of Darcy's law for the volumetric flow rates in two-phase flow takes the form

$$v_\alpha = -\lambda_\alpha(s)\, \nabla\psi_\alpha, \quad \alpha = o, w, \tag{2.1}$$

where the phase mobilities are defined by

$$\lambda_\alpha(s) = \mu_\alpha^{-1}\mathbf{k}k_{r\alpha}(s), \quad \alpha = o, w.$$

Incompressibility and conservation of mass (or, equivalently, volume) imply that

$$\phi\, s_t + \nabla \cdot v_w = q_{ext,w}, \tag{2.2a}$$
$$-\phi\, s_t + \nabla \cdot v_o = q_{ext,o}, \tag{2.2b}$$

where $q_{ext,\alpha}$ is the external volumetric α-source.

Define a "capillary potential" by

$$\psi_c = \psi_o - \psi_w = p_c(s) - (\rho_o - \rho_w)gz, \tag{2.3}$$

Symbol	Meaning	Fractures	Matrix
α	phase: o (oil) or w (water)		
	pressure of α phase	$P_\alpha^\epsilon(x,t)$	$p_\alpha^\epsilon(x,t)$
	saturation of α phase	$S_\alpha^\epsilon(x,t)$	$s_\alpha^\epsilon(x,t)$
	saturation of w phase	$S^\epsilon(x,t)$	$s^\epsilon(x,t)$
	initial saturation	$S_{init}(x)$	$s_{init}(x)$
	porosity (bounded below)	$\Phi^*(x)$	$\phi^\epsilon(x) = \phi(\frac{x}{\epsilon})$ (periodic)
	permeability (bounded below and positive definite)	$\mathbf{K}^*(x)$	$\mathbf{k}^\epsilon(x) = \mathbf{k}(\frac{x}{\epsilon})$ (periodic tensor)
	relative α permeability	$K_{r\alpha}(S^\epsilon)$	$k_{r\alpha}(s^\epsilon)$
	capillary pressure	$P_c(S^\epsilon)$	$p_c(s^\epsilon)$
	residual α saturation	$S_{r\alpha}$	$s_{r\alpha}$
ρ_α	density of α phase		
μ_α	viscosity of α phase		
g	gravitational constant		
$q_{ext,\alpha}(x,t)$	external α source/sink		
	mobility of α phase	$\Lambda_\alpha^*(x,S^\epsilon) = \dfrac{\mathbf{K}^*(x)K_{r\alpha}(S^\epsilon)}{\mu_\alpha}$	$\lambda_\alpha^\epsilon(x,s^\epsilon) = \lambda_\alpha(\frac{x}{\epsilon},s^\epsilon) = \dfrac{\mathbf{k}(\frac{x}{\epsilon})k_{r\alpha}(s^\epsilon)}{\mu_\theta}$

Table 1: Symbols for Immiscible Flow.

and use it and ψ_w as the primary dependent variables. Then,

$$s = p_c^{-1}(\psi_c + (\rho_o - \rho_w)gz) \tag{2.4}$$

is defined from ψ_c and

$$v_o = -\lambda_o(s)\nabla(\psi_w + \psi_c). \tag{2.5}$$

Finally, define a total volumetric flow rate and a total mobility by

$$v = v_o + v_w = -\lambda(s)\,\nabla\psi_w - \lambda_o(s)\nabla\psi_c, \tag{2.6a}$$

$$\lambda(s) = \lambda_w(s) + \lambda_o(s), \tag{2.6b}$$

respectively, and add (2.2a) and (2.2b) to obtain a *pressure* equation,

$$\nabla\cdot v = q_{ext}, \tag{2.7}$$

where $q_{ext} = q_{ext,o} + q_{ext,w}$ is the total external volumetric source. Then, (2.2a) is called the *saturation* equation.

In summary, the basic equations for describing two-phase, incompressible, immiscible flow in a single porosity system are given by (2.1), (2.2a), (2.4), and either (2.2b) and (2.5) or (2.6)-(2.7).

The ϵ-micromodel consists of properly scaled versions of the above equations in the ϵ-reservoir. Physical properties will be discontinuous across Γ^ϵ. Some of the coefficients in the equations on Ω_m^ϵ will be scaled by appropriate powers of ϵ to conserve flow in an appropriate sense; the scaling is intended to preserve the flow across the union of the interfaces between blocks and fractures within a fixed subset of the reservoir as the number of blocks within that subset increases as ϵ tends to zero. The essential physical feature of flow in a fractured reservoir, that the flow equations must reflect a history dependence of the same nature as is seen in viscoelasticity, results from this scaling; see [1] for an explicit reduction of a model for single phase flow in a fractured reservoir to an integro-differential system that illustrates the history dependence in a precise manner; see also [17]. This scaling technique has been used in a number of papers other than ones involving the current authors and their colleagues (see [11], [10], [13], [14], [12], [15], [18] for examples) and allows us to derive a *double* porosity model rather than simply a modified single porosity model which would be inadequate to describe the complexity of flow in a naturally fractured reservoir.

To reduce the number of subscripts in the notation, we use capital letters to indicate quantities in the fractures and small letters to indicate those in the matrix blocks, as indicated in Table 1.

Let $\sigma_w = 1$ and $\sigma_o = -1$. Then, for $\alpha = o, w$, the ϵ–micromodel is given by

$$\sigma_\alpha \Phi^* S_t^\epsilon - \nabla \cdot [\Lambda_\alpha^*(S^\epsilon)\nabla\Psi_\alpha^\epsilon] = q_{ext,\alpha} \quad \text{for} \quad x \in \Omega_f^\epsilon,\ t > 0, \tag{2.8a}$$

$$\sigma_\alpha \phi s_t^\epsilon - \nabla \cdot [\epsilon^2 \lambda_\alpha(s^\epsilon)\nabla\psi_\alpha^\epsilon] = q_{ext,\alpha} \quad \text{for} \quad x \in \Omega_m^\epsilon,\ t > 0, \tag{2.8b}$$

$$\Lambda_\alpha^*(S^\epsilon)\nabla\Psi_\alpha^\epsilon \cdot \nu = \epsilon^2 \lambda_\alpha^\epsilon(s^\epsilon)\nabla\psi_\alpha^\epsilon \cdot \nu \quad \text{for} \quad x \in \Gamma^\epsilon,\ t > 0, \tag{2.8c}$$

$$\Psi_\alpha^\epsilon = \psi_\alpha^\epsilon \quad \text{for} \quad x \in \Gamma^\epsilon,\ t > 0, \tag{2.8d}$$

$$S^\epsilon = S_{init} \quad \text{for} \quad x \in \Omega_f^\epsilon,\ t = 0, \tag{2.8e}$$

$$s^\epsilon = s_{init} \quad \text{for} \quad x \in \Omega_m^\epsilon,\ t = 0, \tag{2.8f}$$

where the solutions are assumed to have the asymptotic forms

$$\begin{aligned} S^\epsilon &= S^0(x,y,t) + \epsilon S^1(x,y,t) + \dots, \\ s^\epsilon &= s^0(x,y,t) + \epsilon s^1(x,y,t) + \dots, \\ \Psi_\alpha^\epsilon &= \Psi_\alpha^0(x,y,t) + \epsilon \Psi_\alpha^1(x,y,t) + \dots, \\ \psi_\alpha^\epsilon &= \psi_\alpha^0(x,y,t) + \epsilon \psi_\alpha^1(x,y,t) + \dots . \end{aligned}$$

The macroscopic and microscopic scales are related by ϵ, that is, up to a translation,

$$y \sim \epsilon^{-1} x,$$

so that

$$\nabla \sim \epsilon^{-1}\nabla_y + \nabla_x,$$

where ∇_z is the gradient with respect to the z–variables. Accordingly, Λ_α^*, λ_α, Ψ_c, ψ_c, and $q_{ext,\alpha}$ are also expanded in powers of ϵ.

We are led to the following formal relations. From (2.8a), in $\Omega \times \mathcal{Q}_f$, the ϵ^{-2}, ϵ^{-1}, and ϵ^0 terms yield

$$-\nabla_y \cdot [\Lambda_\alpha^*(S^0)\nabla_y \Psi_\alpha^0] = 0, \tag{2.9a}$$

$$-\nabla_y \cdot [\Lambda_\alpha^*(S^0)(\nabla_y \Psi_\alpha^1 + \nabla_x \Psi_\alpha^0) + \Lambda_\alpha^1 \nabla_y \Psi_\alpha^0] - \nabla_x \cdot [\Lambda_\alpha^*(S^0)\nabla_y \Psi_\alpha^0] = 0, \quad (2.9b)$$

$$\sigma_\alpha \Phi^* S_t^0 - \nabla_y \cdot [\Lambda_\alpha^*(S^0)(\nabla_y \Psi_\alpha^2 + \nabla_x \Psi_\alpha^1) + \Lambda_\alpha^1(\nabla_y \Psi_\alpha^1 + \nabla_x \Psi_\alpha^0) + \Lambda_\alpha^2 \nabla_y \Psi_\alpha^0]$$
$$-\nabla_x \cdot [\Lambda_\alpha^*(S^0)(\nabla_y \Psi_\alpha^1 + \nabla_x \Psi_\alpha^0) + \Lambda_\alpha^1(S^0)\nabla_y \Psi_\alpha^0] = q_{ext,\alpha}(S^0). \quad (2.9c)$$

From (2.8*c*), on $\Omega \times \partial \mathcal{Q}_m$, the ϵ^{-1}, ϵ^0, and ϵ^1 terms give

$$\Lambda_\alpha^*(S^0)\nabla_y \Psi_\alpha^0 \cdot \nu = 0, \quad (2.10a)$$

$$\Lambda_\alpha^*(S^0)(\nabla_y \Psi_\alpha^1 + \nabla_x \Psi_\alpha^0) \cdot \nu + \Lambda_\alpha^1 \nabla_y \Psi_\alpha^0 \cdot \nu = 0, \quad (2.10b)$$

$$\Lambda_\alpha^*(S^0)(\nabla_y \Psi_\alpha^2 + \nabla_x \Psi_\alpha^1) \cdot \nu + \Lambda_\alpha^1(\nabla_y \Psi_\alpha^1 + \nabla_x \Psi_\alpha^0) \cdot \nu + \Lambda_\alpha^2 \nabla_y \Psi_\alpha^0 \cdot \nu$$
$$= \lambda_\alpha(y, s^0)(\nabla_y \psi_\alpha^0) \cdot \nu. \quad (2.10c)$$

Finally, we only need the ϵ^0 terms from the other equations (2.8*b*), (2.8*d*), (2.8*e*), and (2.8*f*):

$$\begin{aligned}
\phi(y)s_t^0 - \nabla_y \cdot [\lambda_\alpha(y, s^0)(\nabla_y \psi_\alpha^0)] &= q_{ext,\alpha}(s^0) && \text{in} \quad \Omega \times \mathcal{Q}_m,\ t > 0, && (2.11a)\\
\psi_\alpha^0 &= \Psi_\alpha^0 && \text{on} \quad \Omega \times \partial\mathcal{Q}_m,\ t > 0, && (2.11b)\\
S^0 &= S_{init} && \text{on} \quad \Omega \times \mathcal{Q}_f,\ t = 0, && (2.11c)\\
s^0 &= s_{init} && \text{on} \quad \Omega \times \mathcal{Q}_m,\ t = 0. && (2.11d)
\end{aligned}$$

From (2.3),

$$\begin{aligned}
\Psi_o^0 - \Psi_w^0 &= \Psi_c(S^0) \quad \text{in} \quad \Omega \times \mathcal{Q}_f,\ t > 0,\\
\psi_o^0 - \psi_w^0 &= \psi_c(s^0) \quad \text{in} \quad \Omega \times \mathcal{Q}_m,\ t > 0.
\end{aligned}$$

We can show [2] that S^0 and the Ψ_α^0 are independent of y. That is,

$$S^0 = S^0(x,t), \qquad \Psi_\alpha^0 = \Psi_\alpha^0(x,t).$$

Next, (2.9*b*) and (2.10*b*) allow us to write Ψ_α^1 in terms of Ψ_α^0 as

$$\Psi_\alpha^1 = \sum_{j=1}^{3} \omega_j(y)[\partial_j \Psi_\alpha^0(x,t)] + \theta_\alpha(x,t), \quad (2.12)$$

where the ω_j are $\mathcal{Q}$–periodic solutions to the problems

$$\begin{aligned}
\nabla_y^2 \omega_j &= 0 && \text{in} \quad \mathcal{Q}_f, && (2.13a)\\
\nabla_y \omega_j \cdot \nu &= -e_j \cdot \nu && \text{on} \quad \partial\mathcal{Q}_m, && (2.13b)
\end{aligned}$$

$\partial_i = \partial/\partial x_i$, and the θ_α are independent of y.

Finally, locally averaging (2.9*c*) for each α leads us to a macroscopic equation that represents mass conservation and Darcy's law:

$$\Phi S_t^0 - |\mathcal{Q}|^{-1} \int_{\mathcal{Q}_f} \nabla_y \cdot [\Lambda_\alpha^*(S^0)(\nabla_y \Psi_\alpha^2 + \nabla_x \Psi_\alpha^1) + \Lambda_\alpha^1(\nabla_y \Psi_\alpha^1 + \nabla_x \Psi_\alpha^0)]dy$$
$$-|\mathcal{Q}|^{-1} \int_{\mathcal{Q}_f} \nabla_x \cdot [\Lambda_\alpha^*(S^0)(\nabla_y \Psi_\alpha^1 + \nabla_x \Psi_\alpha^0)]dy = |\mathcal{Q}_f|\,|\mathcal{Q}|^{-1} q_{ext,\alpha}(S^0), \quad (2.14)$$

where

$$\Phi = \Phi(x) = |\mathcal{Q}_f|\,|\mathcal{Q}|^{-1}\,\Phi^*.$$

Applying the divergence theorem and making use of equation (2.10*c*) allows the first integral above to be written as

$$\int_{\mathcal{Q}_m} [-\phi(y)s_t^0 + q_{ext,\alpha}(s^0)]dy.$$

Now with (2.12) and (2.13*b*) being used to rewrite the Ψ_α^1 term in the second integral, (2.14) becomes our macroscopic equation

$$\begin{aligned} \Phi S_t^0 + |\mathcal{Q}|^{-1} \int_{\mathcal{Q}_m} \phi(y)s_t^0 dy - \nabla \cdot [\Lambda_\alpha(S^0)(\nabla\Psi_\alpha^0)] \\ = |\mathcal{Q}|^{-1}\Big\{|\mathcal{Q}_f| q_{ext,\alpha}(S^0) + \int_{\mathcal{Q}_m} q_{ext,\alpha}(s^0)dy\Big\} \quad \text{in } \Omega. \end{aligned} \tag{2.15}$$

Here we define the macroscopic fracture system permeability $\mathbf{K}(x)$ as the tensor

$$\mathbf{K} = K^*\Big[|\mathcal{Q}_f|\,|\mathcal{Q}|^{-1}\,\mathbf{I} + \overline{\partial_i\omega_j}\Big], \tag{2.16}$$

where $\mathbf{I}$ is the identity tensor, the double subscripted symbol ($\overline{\partial_i\omega_j}$) is the tensor whose (i,j)–component is as written, and the overbar denotes the local average

$$\overline{\varphi} = |\mathcal{Q}|^{-1} \int_{\mathcal{Q}_f} \varphi(y)dy.$$

Finally, set

$$\Lambda_\alpha(S^0) = \mu_\alpha^{-1}\mathbf{K}K_{r,\alpha}(S^0). \tag{2.17}$$

3 A MODIFIED MEDIUM BLOCK MODEL

In the medium block model described above, one matrix block is associated with each point of the fracture domain; for the example of waterflooding a vertical cross-section, see Figure 3(a). This model can be generalized so as to allow a collection of blocks of different sizes and physical properties to be associated with each point of the reservoir. Let N be the maximum number of distinct block types over any point in the reservoir; let

$$V_i(x) \geq 0, \quad i = 1, \ldots, N; \qquad \sum_{i=1}^N V_i(x) = 1;$$

and let the block physical properties depend on x, y, and i. Also, let the physical parameters in the fractures depend on x. Then, the generalized model requires the incorporation of up to N systems analogous to (2.11) over each point in the fracture domain; let $s_\alpha^i(x,y,t)$, $i = 1, \ldots, N$, denote these block saturations; clearly, only those corresponding to positive $V_i(x)$ need be evaluated. Then, (2.15) must be modified as follows (with the explicit dependence on x being suppressed):

$$\begin{aligned} \Phi S_t + \sum_{i=1}^N V_i|\mathcal{Q}_i|^{-1} \int_{\mathcal{Q}_{i,m}} \phi_i(y)s_t^i dy - \nabla \cdot [\Lambda_\alpha(S)(\nabla\Psi_\alpha)] \\ = \sum_{i=1}^N V_i|\mathcal{Q}_i|^{-1}\Big\{|\mathcal{Q}_{i,f}| q_{ext,\alpha}(S) + \int_{\mathcal{Q}_{i,m}} q_{i,ext,\alpha}(s^i)dy\Big\} \quad \text{in } \Omega. \end{aligned} \tag{3.1}$$

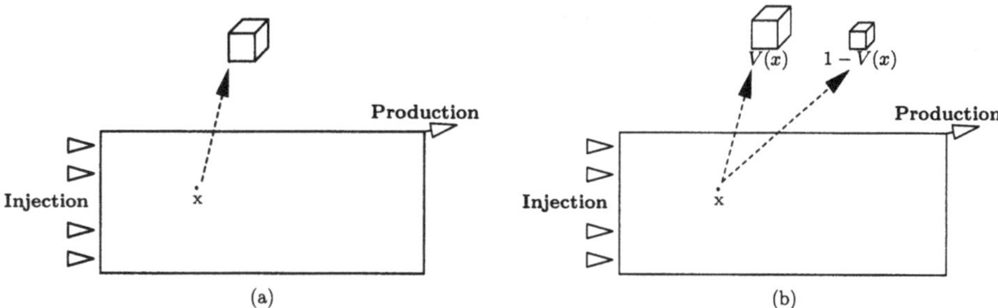

Figure 3: Vertical Cross-Section for the Original and Modified Models

In Figure 3(b), we illustrate this with the simple case where there are two matrix blocks of different sizes associated with a given fracture domain point; the volume fraction of the first block at the point x is $V_1(x) = V(x)$, while the other block has a volume fraction of $V_2(x) = 1 - V(x)$.

4 NUMERICAL EXPERIMENTS

In this section, we will present some numerical results for both vertical cross-section and "five-spot" experiments.

4.1 Vertical Cross-Section Experiments

In these experiments, we used the extended medium block model to allow a distribution of blocks of varying sizes at each point of the domain. The geometry of the reservoir is described as follows: the reservoir is a vertical rectangular cross-section where uniformity is assumed in the third direction, so that the fracture calculations are two-dimensional while the block calculations are three-dimensional. The height of the reservoir is 10 meters and its length is 300 meters.

The reservoir is assumed to be initially in gravitational and capillary equilibrium with an average water saturation of 25%. Injection occurs at a uniform rate of .2 pore volumes per year along the left face of the reservoir. Oil and water are produced in proportion to their mobilities at the top right corner.

The capillary pressure functions were taken to be

$$P_c(S) = (1-S)[\gamma(S^{-1}-1)+\Theta], \tag{4.1}$$

$$p_c(s) = \delta\Big(\frac{1}{(s-s_{rw})^2} - \frac{s_{ro}^2}{(1-s)^2(1-s_{ro}-s_{rw})^2}\Big), \tag{4.2}$$

and the relative permeabilities were taken to be

$$K_{ro}(S) = 1 - S, \qquad K_{rw}(S) = S, \tag{4.3}$$

$$k_{ro}(s) = \Big(1 - \frac{s}{1-s_{ro}}\Big)^2, \qquad k_{rw}(s) = \Big(\frac{s-s_{rw}}{1-s_{rw}}\Big)^2. \tag{4.4}$$

The values which were used for other physical parameters are given in Table 2. The absolute permeability tensors were assumed to be diagonal: $\mathbf{K} = K\mathbf{I}$ and $\mathbf{k} = k\mathbf{I}$ in all of the experiments.

parameter	value	parameter	value
μ_w	.5 cp	μ_o	2 cp
ρ_w	1 g/cm^3	ρ_o	.7 g/cm^3
K	1 darcy	k	0.005 darcy
Φ	.01	ϕ	.2
s_{ro}	.15	s_{rw}	.2
γ	20,000 dynes/cm^2	Θ	100 dynes/cm^2
δ	1,500 dynes/cm^2		

Table 2: Default Parameter Values.

Other numerical experiments have shown that a relatively coarse discretization of both the fracture and blocks produces an adequate solution. The discretization used for the fracture system was 40 nodes in the horizontal direction and 10 in the vertical direction; 16 internal grid points were used in the discretization of each matrix block.

To examine the effects of varying the block size or the distribution of the matrix blocks we present some production curves. Production curves show the total oil produced versus the total water injected and give a good idea of the global behavior of the model. In Figure 4, production curves are shown for several different cases. For reference, curves are shown for the cases where the reservoir is homogeneous and consists entirely of blocks 50 *cm* in diameter and for the case where all the blocks are 200 *cm* in diameter. Two layered reservoirs are also represented: one in which the top half of the reservoir consists of 50 *cm* blocks and the bottom half consists of 200 *cm* blocks and the other where the layers are reversed. In addition, we show the results of using two different sizes of blocks (50 *cm* and 200 *cm*) at each point. This case is labeled "composite".

These production curves show some features worth mentioning. First, one finds less oil production for larger blocks. This is logical considering that we have a constant fracture porosity. Capillary forces have a much more difficult time in forcing oil out of the larger blocks; thus, water flows around the blocks and arrives at the production well. Second, the order of the layering is important. This is due to gravitational segregation of the fluids. When the larger blocks are on the bottom of the reservoir, the water which falls to the bottom is not imbibed by the blocks as rapidly. This water then remains in the fracture network until it finds its way to the production well.

Analagous curves are shown in Figure 5 where blocks of size 400 *cm* were used in place of 200 *cm* blocks. The results are similar to those seen in Figure 4 but are more pronounced due to the larger variation in block size.

In Figure 6, production curves are shown for the composite case with 50 *cm* and 200 *cm* blocks at each fracture point and for a case where there is one block associated with each fracture grid point but its size is randomly chosen to be either 50 *cm* or 200 *cm*. The two curves are virtually identical, indicating that there is little difference in the global fluid flow between the two cases.

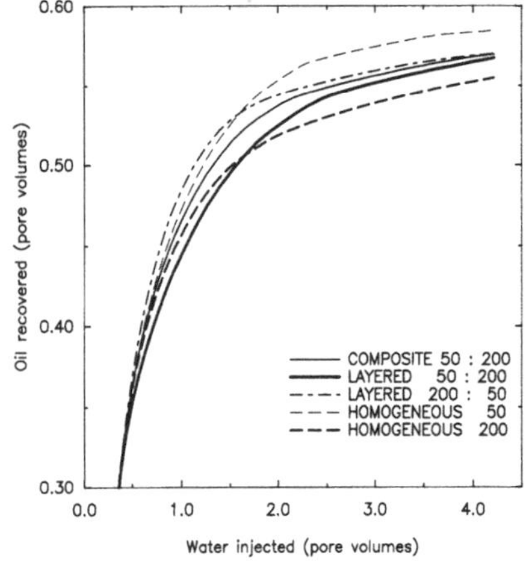

Figure 4: Production curves for various configurations and block sizes with 50 cm and 200 cm blocks.

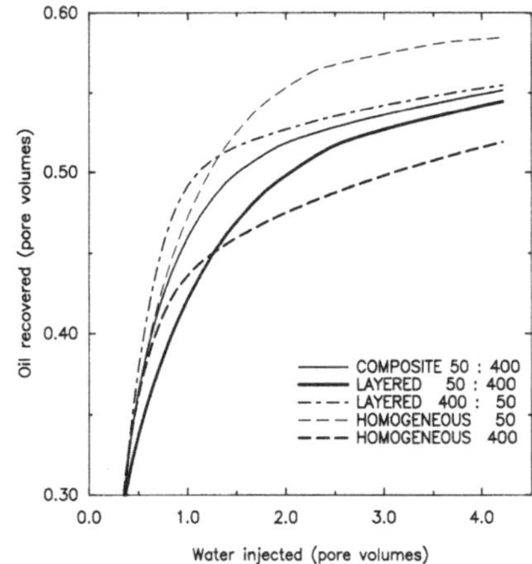

Figure 5: Production curves for various configurations and block sizes with 50 cm and 400 cm blocks.

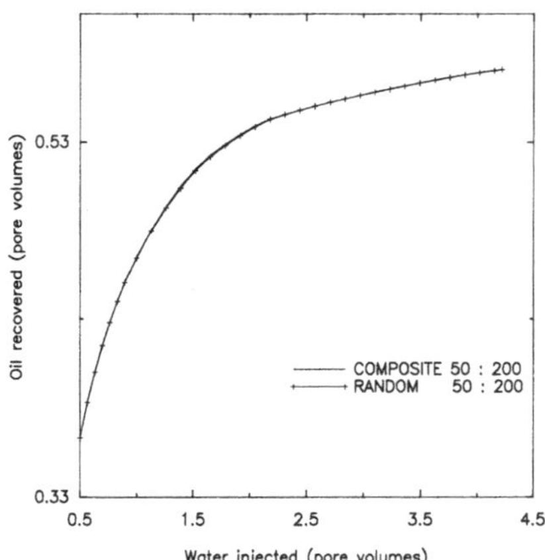

Figure 6: Production curves comparing composite block with randomly distributed blocks.

parameter	value	parameter	value
μ_w	.5 cp	μ_o	2 cp
ρ_w	1 g/cm^3	ρ_o	.7 g/cm^3
s_{ro}	.15	s_{rw}	.2
γ	2.0 dynes/cm^2	Θ	0.01 dynes/cm^2
δ	0.3 dynes/cm^2		

Table 3: Parameter Values for 5-spot experiments.

4.2 Five spot experiments

In these experiments, we consider a horizontal section of a reservoir with length and width of 300 m. Injection occurs at the lower left hand corner of the domain and production at the upper right corner. The discretization used in the fracture system was 40 nodes in both spatial directions. The matrix blocks are now considered to be two-dimensional (gravity is ignored). As before, the reservoir is assumed to be initially in capillary equilibrium with an average water saturation of 25%. Water is injected at a uniform rate of .2 pore volumes per year.

Our goal is to examine the effects which inhomogeneities in both the porosity and permeability in the matrix blocks and the fracture system have on the fluid flow. Table 3 gives the physical properties which are common to all of the experiments which are described below.

To see the effects of the heterogeneities, we present contour plots of both the water saturation in the fractures and the average matrix water saturation. In Figures 7 and 8, we have the contours after 1200 days for a reference homogeneous case where the matrix porosity is $\phi = .2$, the matrix permeability is $k = 0.005\, darcy$, and the fracture permeability is $K = 1\, darcy$. The fracture porosity is $\Phi = .01$ and will remain fixed in all the experiments. The matrix blocks have a diameter of 50 cm.

In Figure 9 and 10, we show saturation contours where the matrix porosity and permeability, as well as the fracture permeability, vary. For a given block, the porosity and permeability were randomly chosen such that $.04 \leq \phi \leq .4$, and k is between .001 and .01 $darcy$. The fracture permeability at a given grid point was randomly chosen to be either .5 $darcy$ or 1.5 $darcy$.

The same experiment was run again with a different sequence of random numbers. The results are shown in Figures 11 and 12.

In Figures 13 and 14, the matrix porosity and permeability were randomly varied as above, but the fracture permeability was held constant ($K = 1\, darcy$) throughout the reservoir. In Figures 15 and 16, the matrix porosity and permeability were held constant ($\phi = .2$ and $k = .01\, darcy$) and the fracture permeability was randomly chosen to be either .5 or 1.5 $darcy$.

The saturation contours clearly show that large variations in the fracture permeability are more likely to affect the profile of the front. This, of course, is quite reasonable. The variations in the matrix properties cause only minor changes in the saturation contours. It is also reasonable to note that the random variations in the physical properties does not greatly affect the average fluid flow. Figure 17 shows production curves

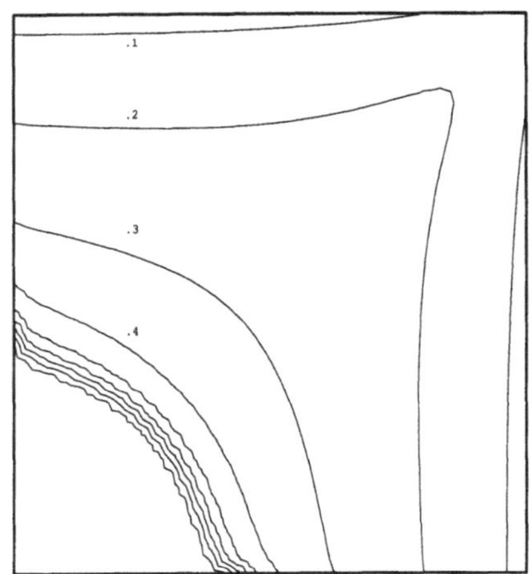

Figure 7: Fracture water saturation contours for homogeneous reservoir with 50 cm blocks after 1200 days.

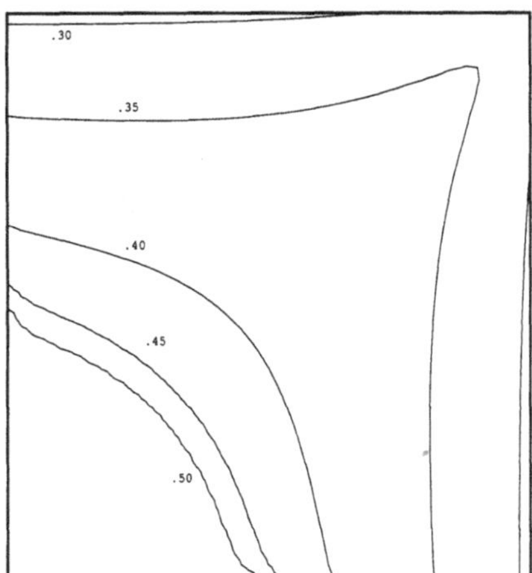

Figure 8: Average matrix water saturation contours for homogeneous reservoir with 50 cm blocks after 1200 days.

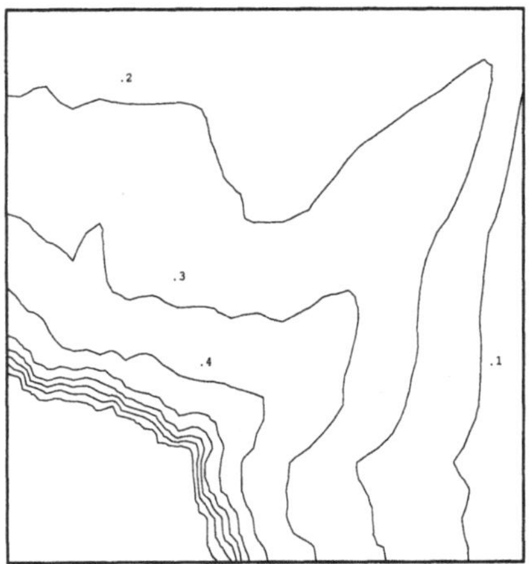

Figure 9: Fracture water saturation contours with random variations in both the matrix permeability and porosity and the fracture permeability.

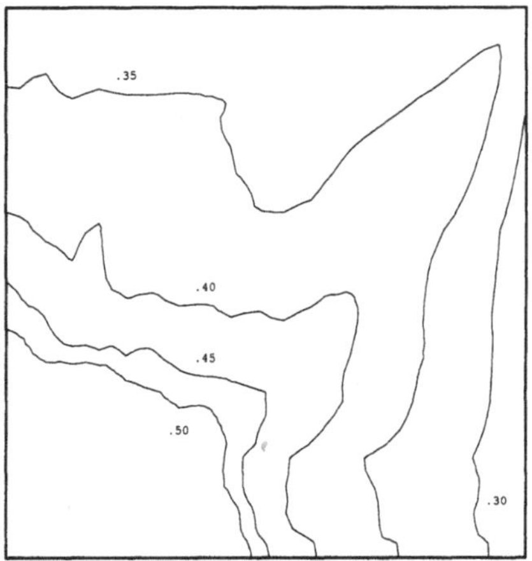

Figure 10: Average matrix water saturation contours with random variations in both the matrix permeability and porosity and the fracture permeability.

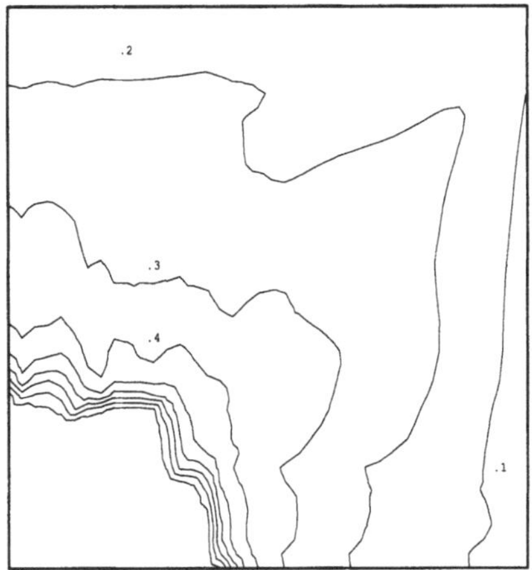

Figure 11: Fracture water saturation contours with random variations in both the matrix permeability and porosity and the fracture permeability.

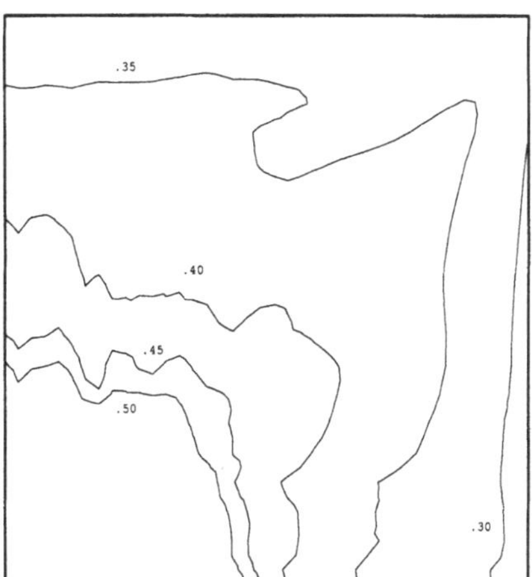

Figure 12: Average matrix saturation contours with random variations in both the matrix permeability and porosity and the fracture permeability.

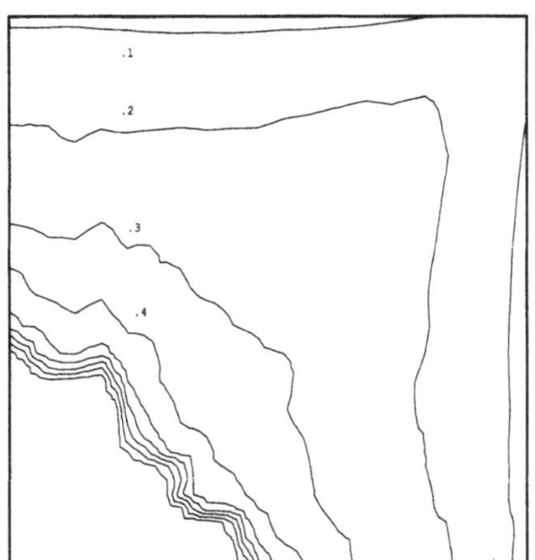

Figure 13: Fracture water saturation contours with random variations in the matrix permeability and porosity and constant fracture permeability.

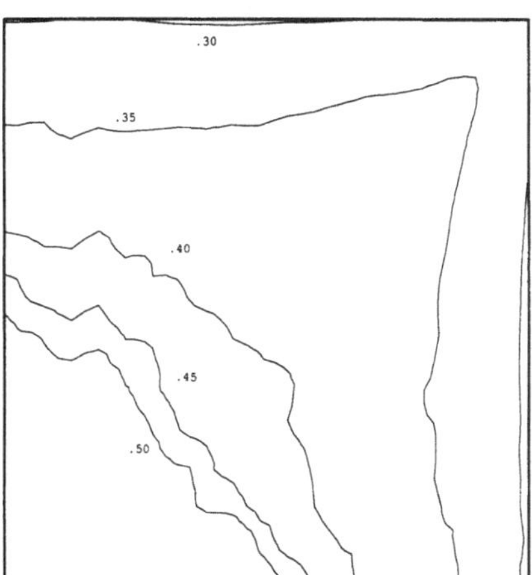

Figure 14: Average matrix water saturation contours with random variations in the matrix permeability and porosity and constant fracture permeability.

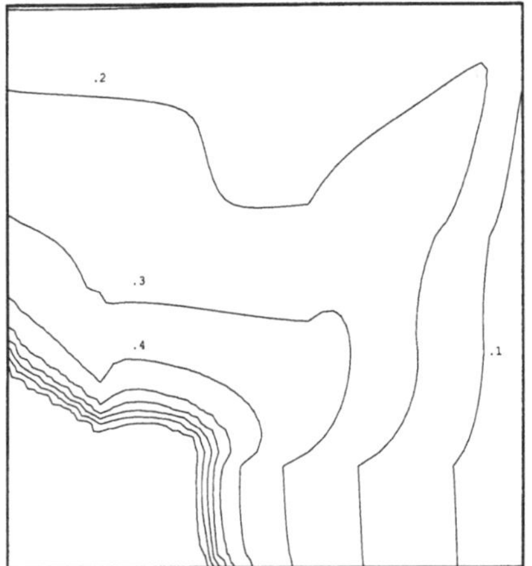

Figure 15: Fracture water saturation contours with constant matrix permeability and porosity and random variations in the fracture permeability.

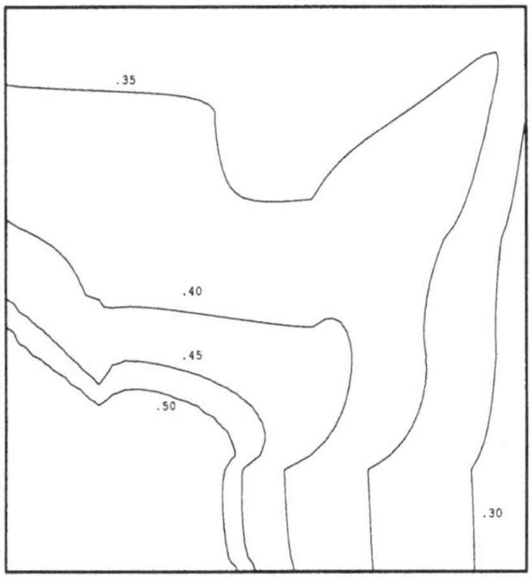

Figure 16: Average matrix water saturation contours with constant matrix permeability and porosity and random variations in the fracture permeability.

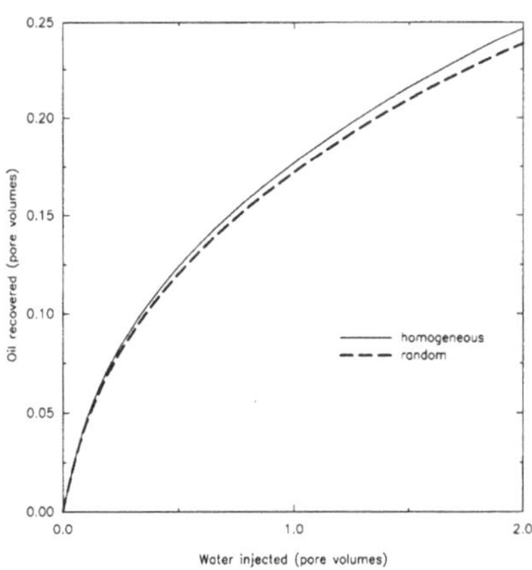

Figure 17: Production curves comparing the homogeneous reservoir with one with random variations in both matrix and fracture properties.

for two cases: the homogeneous reservoir and the case where both matrix and fracture properties were varied (see Figures 9 and 10). The difference in oil production is caused by a slight difference in the effective average fracture permeability.

5 CONCLUSIONS

The model for immiscible, incompressible flow in naturally fractured petroleum reservoirs previously discussed in [2], [7], and [9] has been successfully generalized to allow a reasonably realistic treatment of inhomogeneities in both the fractures and the matrix blocks. In the case of a vertical cross-section experiment, as a result of gravity and imbibition, inhomogeneities had a noticeable effect on recovery curves; i.e., the local changes in the flow pattern lead to global changes. In the five-spot experiments, inhomogeneities caused significant changes in the saturation patterns, but the recovery curves were not seriously modified. These preliminary calculations indicate the need to continue this study with a full three-dimensional treatment of the fractures as well as the blocks; such a study is currently being pursued.

REFERENCES

[1] Arbogast T. The double porosity model for single phase flow in naturally fractured reservoirs. In *Numerical Simulation in Oil Recovery*, The IMA Volumes in Mathematics and its Applications 11, pages 23–45. Springer-Verlag, Berlin and New York, 1988. M. F. Wheeler, ed.

[2] Arbogast T., Douglas J., Hornung U. Modeling of naturally fractured petroleum reservoirs by formal homogenization techniques. In *Frontiers in Pure and Applied Mathematics*, pages 1–19. Elsevier, Amsterdam, 1991. R. Dautray, ed.

[3] Aziz K. and Settari T. *Petroleum Reservoir Simulation.* Applied Science Publishers, London, 1979.

[4] Barenblatt G. I., Zheltov I. P., Kochina I. N. Basic concepts in the theory of seepage of homogeneous liquids in fissured rocks. *J. Appl. Math. and Mech.*, 24:1286–1303, 1960.

[5] Chavent G., Jaffré J. *Mathematical Models and Finite Elements for Reservoir Simulation.* North-Holland, Amsterdam, 1986.

[6] de Swaan A. Theory of waterflooding in fractured reservoirs. *Soc. Petroleum Engr. J.*, 18:117–122, 1978.

[7] Douglas J., Jr., Arbogast T. Dual porosity models for flow in naturally fractured reservoirs. In *Dynamics of Fluids in Hierarchical Porous Formations*, pages 177–221. Academic Press, London, 1990. J. H. Cushman, ed.

[8] Douglas J., Jr., Arbogast T., Paes Leme P. J., Hensley J. L., Nunes N. P. Immiscible displacement in vertically fractured reservoirs. *Transport in Porous Media.* To appear, 1993.

[9] Douglas J., Jr., Hensley J. L., Arbogast T. A dual–porosity model for waterflooding in naturally fractured reservoirs. *Computer Methods in Applied Mechanics and Engineering*, 87:157–174, 1991.

[10] Hornung U. Applications of the homogenization method to flow and transport through porous media. In *Flow and Transport in Porous Media*, Singapore. Summer School, Beijing 1988, World Scientific. Xiao Shutie, ed., to appear.

[11] Hornung U. Miscible displacement in porous media influenced by mobile and immobile water. *Rocky Mtn. Jour. Math.*, 21:645–669, 1991. Corr. pages 1153–1158.

[12] Hornung U., Jäger, W. Homogenization of reactive transport through porous media. In *EQUADIFF 1991*, Singapore. World Scientific Publishing. C. Perelló, ed., submitted 1992.

[13] Hornung U., Jäger W. A model for chemical reactions in porous media. In *Complex Chemical Reaction Systems. Mathematical Modeling and Simulation*, volume 47 of *Chemical Physics*, pages 318–334. Springer, Berlin, 1987. J. Warnatz and W. Jäger, eds.

[14] Hornung U., Jäger W. Diffusion, convection, adsorption, and reaction of chemicals in porous media. *J. Diff. Equations*, 92:199–225, 1991.

[15] Hornung U., Showalter R. E. Diffusion models for fractured media. *Jour. Math. Anal. Appl.*, 147:69–80, 1990.

[16] Peaceman D. W. *Fundamentals of Numerical Reservoir Simulation.* Elsevier, New York, 1977.

[17] Showalter R. E. Distributed microstructure models of porous media. This volume.

[18] Vogt Ch. A homogenization theorem leading to a Volterra integro-differential equation for permeation chromotography. Preprint #155, Sonderfachbereich 123, 1982.

[19] Warren J. E., Root P. J. The behavior of naturally fractured reservoirs. *Soc. Petr. Eng. J.*, 3:245–255, 1963.

International Series of Numerical Mathematics, Vol. 114, © 1993 Birkhäuser Verlag Basel

A Massively Parallel Iterative Numerical Algorithm for Immiscible Flow in Naturally Fractured Reservoirs

Jim Douglas, Jr.* P. J. Paes Leme† Felipe Pereira‡
Li-Ming Yeh‡

Abstract. We propose a new iterative numerical scheme designed for massively parallel processing for an immiscible displacement in a naturally fractured reservoir. The procedure is based on a domain decomposition technique applied to a mixed finite element approximation of the problem; the domain is decomposed into individual elements. Numerical experiments are presented to illustrate its performance on a CM-5 system.

1 INTRODUCTION

High quality numerical simulations of fluid flow in petroleum reservoirs require the use of increasingly finer grids in the numerical discretization of the governing system of partial differential equations so that a large number of length scales relevant to the problem can be incorporated into the simulations. This problem is critical when inhomogeneities are present and their influence need to be adequately resolved.

Detailed two-dimensional studies of the effect of the inhomogeneities of a single porosity medium have reached the limits of existing serial computers [22], [23]. For dual porosity models, meaningful studies are infeasible on serial machines.

In this paper, a parallel iterative procedure, specially designed for massively parallel processing, is proposed for the numerical solution of dual-porosity models for immiscible flow in a naturally fractured reservoir. Two implementations were performed, one completely portable, adequate for MIMD systems, and the other using a data-parallel programing language particular to a Connection Machine Model CM-5 using SIMD control. For fluid flow simulations with grid sizes relevant for applications we find a good scalability of our algorithm, with a consistent slightly better performance of the MIMD version. However, through the use of vector processing units, the SIMD code runs faster.

The model problem treated in this paper corresponds physically to a waterflooding of a naturally fractured petroleum reservoir where the average spacing between fractures is relatively small compared to the reservoir size. With the terminology adopted in previous works the particular model treated herein is known as the "medium block model" [16]. The system of partial differential equations governing fluid flow in this double porosity formulation treats the flow in each matrix block in a completely parallelizable fashion. The part of the system describing the flow in the fractures is then

*Department of Mathematics, Purdue University, West Lafayette, IN 47907-1395
†Instituto Politécnico, Universidade do Estado do Rio de Janeiro, Nova Friburgo, Brazil
‡Department of Mathematics, Purdue University, West Lafayette, IN 47907-1395

numerically approximated by a hybridized mixed finite element method. A domain decomposition technique in which the domain is decomposed into individual elements is then applied. This allows us to adapt the solution of the problem to massively parallel processing. Domain decomposition techniques distinct from the one used here can be found in [7], [24], [25], [19], and [18]. As our first step towards full three-dimensional fluid flow simulations, we consider in the numerical studies described in this work two-dimensional fractured media to which are attached three-dimensional matrix blocks.

We define a nonlinear iterative procedure and use it to solve numerically the part of the system describing flow in the fractures which is coupled to the system of equations for the matrix blocks through source terms. This method is motivated by the linear problem analyzed in [17], which is closely related to the one introduced in [9] for a Helmholz problem and extended to another Helmholz–like problem related to Maxwell's equations [8], [10]. As in the above references, we shall make use of the hybridization of mixed finite element methods introduced in [20] and [21] more than twenty–five years ago and which has been carefully analyzed in [1]; see also [4], [2], and [3]. For the numerical solution of the local problems associated with the matrix blocks, a simple finite difference scheme [16] will be used. A rigorous proof of convergence of the iterative procedure is currently being investigated by the authors. For the simpler problem of two-phase flow through a single porosity medium, convergence of the iteration has been established.

This paper is organized in the following way. In §2 a brief description of the model considered here, along with a time discretization for it, is given. A domain decomposition technique and the new iterative procedure defined for a mixed finite element approximation of the system of equations in the fractures appear in §3. A detailed description of the time-dependent algorithm developed to solve the full governing system appears in §4. Distinct parallel implementations of our numerical method in a CM-5 system are analyzed in §5. Finally, §6 is devoted to our conclusions and interesting open problems related to this work.

2 THE MODEL PROBLEM

2.1 Governing Equations

We consider saturated, two-phase, incompressible, immiscible flow, the phases being o (oil or nonwetting phase) and w (water or wetting phase), with densities and viscosities ρ_α and μ_α, $\alpha = o, w$, respectively. See [15] in this volume for a description of the system of equations governing fluid flow in a single porosity model under the above assumptions.

The governing system for the medium block model is derived through the mathematical theory of homogenization [12]. It produces a two-phase, single-porosity model for the flow in the matrix system and a second, slightly modified single-porosity system in the fractures. To reduce the number of subscripts in the notation, we use capital letters to indicate quantities in the fractures and small letters to indicate those in the matrix blocks.

Let Ω_x denote the block attached to the point $x \in \Omega$; the w-saturation in Ω_x will be indicated by $s(x, y, t)$, $x \in \Omega$, $y \in \Omega_x$, $t \geq 0$, etc.

The capillary pressure and relative permeability functions are somewhat different in the fractures than in the matrix blocks. Generally, one assumes that the fractures are essentially like spaces between two parallel planes and that $S_{\min} = 0$ and $S_{\max} = 1$. The singularity in the capillary pressure curve as S decreases to $S_{\min} = 0$ is weaker than that for the capillary pressure function in the blocks, and the relative permeability functions can be taken to be linear or nearly linear. The absolute permeability tensor on the fracture sheet reflects the geometry of the blocks [11].

The source terms in the saturation and pressure equations in the fractures contain two terms, one defining the external flow (wells in practice). In addition, there are matrix source terms $q_{m,\,\alpha}$, $\alpha = o, w$, for each of the phases. The system governing flow in the fracture system can be written as

$$\Phi\frac{\partial S}{\partial t} - \nabla\cdot\mathbf{Q}_w = q_{\text{ext},w} + q_{m,w} \quad \text{for } x\in\Omega,\ t>0, \tag{2.1a}$$

$$\mathbf{Q}_w = -\tilde{\Lambda}_w(S)\nabla\Psi_w \quad \text{for } x\in\Omega,\ t>0, \tag{2.1b}$$

$$-\Phi\frac{\partial S}{\partial t} - \nabla\cdot\mathbf{Q}_o = q_{\text{ext},o} + q_{m,o} \quad \text{for } x\in\Omega,\ t>0, \tag{2.1c}$$

$$\mathbf{Q}_o = -\tilde{\Lambda}_o(S)\nabla\Psi_o \quad \text{for } x\in\Omega,\ t>0, \tag{2.1d}$$

$$S = P_c^{-1}(\Psi_c + (\rho_o - \rho_w)gz), \tag{2.1e}$$

Incompressibility requires that $q_{m,o} + q_{m,w} = 0$.

It is convenient to write the equations on the block Ω_x as

$$\phi\frac{\partial s}{\partial t} - \nabla\cdot\left[\tilde{\lambda}_w(s)\nabla\psi_w\right] = 0 \quad \text{for } y\in\Omega_x,\ t>0, \tag{2.2a}$$

$$-\nabla\cdot\left[\tilde{\lambda}(s)\nabla\psi_w + \tilde{\lambda}_o(s)\nabla\psi_c\right] = 0 \quad \text{for } y\in\Omega_x,\ t>0, \tag{2.2b}$$

$$s = p_c^{-1}(\psi_c + (\rho_o - \rho_w)gz); \tag{2.2c}$$

we have assumed that the external sources affect the fracture system only. The boundary conditions for the matrix problems are given by requiring continuity of the potentials:

$$\psi_w(x,y,t) = \Psi_w(x,t) \quad \text{for } y\in\partial\Omega_x,\ x\in\Omega,\ t>0, \tag{2.3a}$$

and

$$\psi_c(x,y,t) = \Psi_c(x,t) \quad \text{for } y\in\partial\Omega_x,\ x\in\Omega,\ t>0. \tag{2.3b}$$

The matrix source terms are defined as follows. The volume of the w-fluid leaving the block Ω_x is

$$\int_{\partial\Omega_x} v_w\cdot n\,da(y) = \int_{\Omega_x}\nabla\cdot v_w\,dy = -\int_{\Omega_x}\phi\frac{\partial s}{\partial t}\,dy;$$

consequently, let

$$q_{m,w}(x,t) = -\frac{1}{|\Omega_x|}\int_{\Omega_x}\phi\frac{\partial s}{\partial t}\,dy \quad \text{for } x\in\Omega,\ t>0. \tag{2.4}$$

We complete the model by specifying the external boundary conditions and the initial conditions for the system. For the case of no flow across the external boundary,

$$\tilde{\Lambda}_\alpha(s)\,\nabla\,\Psi_\alpha\cdot n = 0 \quad \text{for } x\in\partial\Omega,\ t>0,\quad \alpha = o,w. \tag{2.5}$$

Initial saturations (i.e., capillary potentials) must be specified:

$$\begin{aligned}\Psi_c(x,0) &= \Psi_{\text{init},c}(x) \quad \text{for } x \in \Omega,\\ \psi_c(x,y,0) &= \psi_{\text{init},c}(x,y) \quad \text{for } y \in \Omega_x, x \in \Omega.\end{aligned}$$

To be consistent, (2.3) and (2.5) should hold when $t = 0$.

2.2 Time-Discretization

Discretize the time variable by choosing $t^0, t^1, t^2, \ldots, t^N$ such that $0 = t^0 < t^1 < t^2 < \ldots < t^N$, and set $\Delta t^n = t^n - t^{n-1}$. An approximation to a function Θ related to the fracture system at a point $x \in \Omega$ at time t^n will be denoted by

$$\Theta^n \approx \Theta(x, t^n).$$

Approximate (2.1) implicitly by backwards Euler approximations in time to obtain the system

$$\begin{aligned}\Phi\frac{S^n - S^{n-1}}{\Delta t^n} - \nabla\cdot\mathbf{Q}_w^n &= q_{\text{ext},w}^n + q_{m,w}^n \quad \text{for } x \in \Omega,\ t > 0, &(2.6a)\\ \mathbf{Q}_w^n &= -\tilde{\Lambda}_w(S^n)\nabla\Psi_w^n \quad \text{for } x \in \Omega,\ t > 0, &(2.6b)\\ -\Phi\frac{S^n - S^{n-1}}{\Delta t^n} - \nabla\cdot\mathbf{Q}_o^n &= q_{\text{ext},o}^n + q_{m,o}^n \quad \text{for } x \in \Omega,\ t > 0, &(2.6c)\\ \mathbf{Q}_o^n &= -\tilde{\Lambda}_o(S^n)\nabla\Psi_o^n \quad \text{for } x \in \Omega,\ t > 0, &(2.6d)\\ S^n &= P_c^{-1}(\Psi_c^n + (\rho_o - \rho_w)gz), &(2.6e)\end{aligned}$$

The time-discretization of the equations describing the flow in the matrix blocks will be discussed in the context of a finite difference discretization of the matrix system in §4.

3 DOMAIN DECOMPOSITION FOR THE FRACTURE SYSTEM

Parallelization of the solution of the global fracture system problem is achieved through a spatial decomposition, which we now describe.

Let $\Omega \subset \mathbf{R}^d$, $d = 2$ or 3, be a bounded domain with a Lipschitz boundary $\partial\Omega$. Let $\{\Omega_j,\ j = 1, \ldots, M\}$ be a partition of Ω:

$$\overline{\Omega} = \cup_{j=1}^M \overline{\Omega}_j; \quad \Omega_j \cap \Omega_k = \emptyset, \quad j \neq k.$$

Assume that $\partial\Omega_j$, $j = 1, \ldots, M$, is also Lipschitz and that Ω_j is star-shaped. In practice, with the exception of perhaps a few Ω_j's along $\partial\Omega$, each Ω_j would be convex with a piecewise-smooth boundary. Let

$$\Gamma = \partial\Omega, \quad \Gamma_j = \Gamma \cap \partial\Omega_j, \quad \Gamma_{jk} = \Gamma_{kj} = \partial\Omega_j \cap \partial\Omega_k.$$

Let us consider decomposing (2.6) over the partition $\{\Omega_j\}$. In addition to requiring $\{S_j^n\}$, $\{\mathbf{Q}_{\alpha,j}^n\}$, $\{\Psi_{\alpha,j}^n\}$, $\alpha = w, o$, $j = 1, \ldots, M$, to be a solution of (2.6) for $x \in \Omega_j$, $j = 1, \ldots, M$ it is necessary to impose the consistency conditions

$$\begin{aligned}\Psi_{\alpha,j} &= \Psi_{\alpha,k}, \quad & x \in \Gamma_{jk},\ \alpha = w, o, \quad &(3.1a)\\ \mathbf{Q}_{\alpha,j}\cdot\nu_j + \mathbf{Q}_{\alpha,k}\cdot\nu_k &= 0, \quad & x \in \Gamma_{jk},\ \alpha = w, o, \quad &(3.1b)\end{aligned}$$

where ν_j is the unit outer normal to Ω_j.

3.1 Weak Formulation

Let $\mathbf{V}_j = H(\operatorname{div},\Omega_j)$ and $\mathbf{W}_j = L^2(\Omega_j)$ for $j = 1,\dots,M$. The weak formulation of (2.6) with the domain decomposed according to the discussion above is given by seeking $\{S_j^n, \mathbf{Q}_{w,j}^n, \mathbf{Q}_{o,j}^n, \Psi_{w,j}^n, \Psi_{o,j}^n\} \in \mathbf{W}_j \times \mathbf{V}_j \times \mathbf{V}_j \times \mathbf{W}_j \times \mathbf{W}_j$, $j = 1,\dots,M$, such that

$$\frac{(\Phi S_j^n, w_1)_{\Omega_j} - (\Phi S_j^{n-1}, w_1)_{\Omega_j}}{\Delta t^n} - (\nabla\cdot\mathbf{Q}_{w,j}^n, w_1)_{\Omega_j} \tag{3.2a}$$
$$= (q_{\text{ext},w}^n, w_1)_{\Omega_j} + (q_{m,w}^n, w_1)_{\Omega_j},$$

$$\left(\frac{\mathbf{Q}_{w,j}^n}{\tilde{\Lambda}_w(S_j^n)}, \mathbf{v}_1\right)_{\Omega_j} - (\Psi_{w,j}^n, \operatorname{div}\mathbf{v}_1)_{\Omega_j} + \langle\Psi_{w,j}^n, \mathbf{v}_1\cdot\nu\rangle_{\partial\Omega_j} = 0, \tag{3.2b}$$

$$-\frac{(\Phi S_j^n, w_2)_{\Omega_j} - (\Phi S_j^{n-1}, w_2)_{\Omega_j}}{\Delta t^n} - (\nabla\cdot\mathbf{Q}_{o,j}^n, w_2)_{\Omega_j} \tag{3.2c}$$
$$= (q_{\text{ext},o}^n, w_2)_{\Omega_j} + (q_{m,o}^n, w_2)_{\Omega_j},$$

$$\left(\frac{\mathbf{Q}_{o,j}^n}{\tilde{\Lambda}_o(S_j^n)}, \mathbf{v}_2\right)_{\Omega_j} - (\Psi_{o,j}^n, \operatorname{div}\mathbf{v}_2)_{\Omega_j} + \langle\Psi_{o,j}^n, \mathbf{v}_2\cdot\nu\rangle_{\partial\Omega_j} = 0, \tag{3.2d}$$

$$(P_c(S_j^n), w_3)_{\Omega_j} = (\Psi_{c,j}^n, w_3)_{\Omega_j} + ((\rho_o - \rho_w)gz, w_3)_{\Omega_j}, \tag{3.2e}$$

where $\mathbf{v}_1, \mathbf{v}_2 \in \mathbf{V}_j$ and $w_1, w_2, w_3 \in \mathbf{W}_j$. There is a technical difficulty with (3.2*b*) and (3.2*d*); the meaning of the restriction of an L^2–function on Ω_k to Γ_{jk} is not clear. Thus, (3.2) is properly viewed as motivation for the treatment of the discrete case to be discussed below.

3.2 Mixed Finite Element Approximation

We shall treat the case in which $\{\Omega_j\}$ is a partition of Ω into individual elements (simplices, rectangles, prisms), though an inspection of the procedure would indicate that larger subdomains are permissible. Let $\mathbf{W}^h \times \mathbf{V}^h$ be a mixed finite element space over $\{\Omega_j\}$; any of the usual choices is acceptable: [4], [2], [3], [6], [26], [28], [29]. Each of these spaces is defined through local spaces $\mathbf{V}_j^h \times \mathbf{W}_j^h = \mathbf{V}(\Omega_j) \times \mathbf{W}(\Omega_j)$, and setting

$$\begin{aligned}\mathbf{V}^h &= \{\mathbf{v} \in H(\operatorname{div},\Omega) : \mathbf{v}|_{\Omega_j} \in \mathbf{V}_j^h\},\\ \mathbf{W}^h &= \{w : w|_{\Omega_j} \in \mathbf{W}_j^h\}.\end{aligned}$$

In each space $\mathbf{W}^h$ in the various families of mixed elements referenced above, the functions $w \in \mathbf{W}^h$ are allowed to be discontinuous across each Γ_{jk}. As a consequence, attempting to impose the consistency conditions (3.1) would force a flux conservation error; i.e., (3.1*b*) would not be satisfied unless the approximate solution $\Psi_\alpha^h \in \mathbf{W}^h, \alpha = w, o$, to the discrete analogue of (3.1*b*) is constant, a totally uninteresting case. So, let us introduce Lagrange multipliers [20], [21], [1] on the edges $\{\Gamma_{jk}\}$. In the discussion below we consider the parameter α to be either o or w. Assume that, when $\mathbf{Q}_{\alpha,j} = \mathbf{Q}_{\alpha,j}^h|_{\Omega_j}$, $\mathbf{Q}_{\alpha,j}^h \in \mathbf{V}^h$, its normal component $\mathbf{Q}_{\alpha,j}\cdot\nu_j$ on Γ_{jk}, is a polynomial of some fixed degree τ_α, where for simplicity we shall assume τ_α independent of Γ_{jk} (see [5] if not). Set

$$\Lambda_\alpha^h = \{\lambda_\alpha : \lambda_\alpha|_{\Gamma_{jk}} \in P_{\tau_\alpha}(\Gamma_{jk}) = \Lambda_{\alpha,jk}, \Gamma_{jk} \neq \emptyset\};$$

note that there are two copies of P_{τ_α} assigned to the set Γ_{jk}: $\Lambda_{\alpha,jk}$ and $\Lambda_{\alpha,kj}$.

The hybridized mixed finite element method is given by dropping the superscript h and seeking

$$\{S_j^n \in \mathbf{W}_j, \mathbf{Q}_{\alpha,j}^n \in \mathbf{V}_j, \Psi_{\alpha,j}^n \in \mathbf{W}_j, \lambda_{\alpha,jk}^n \in \Lambda_{\alpha,jk}\},$$

where $j = 1, \ldots, M; k = 1, \ldots, M; \alpha = w, o$, such that

$$\frac{(\Phi S_j^n, w_1)_{\Omega_j} - (\Phi S_j^{n-1}, w_1)_{\Omega_j}}{\Delta t^n} - (\nabla \cdot \mathbf{Q}_{w,j}^n, w_1)_{\Omega_j} \tag{3.3a}$$
$$= (q_{\text{ext},w}^n, w_1)_{\Omega_j} + (q_{m,w}^n, w_1)_{\Omega_j},$$

$$\left(\frac{\mathbf{Q}_{w,j}^n}{\tilde{\Lambda}_w(S_j^n)}, \mathbf{v}_1\right)_{\Omega_j} - (\Psi_{w,j}^n, \operatorname{div} \mathbf{v}_1)_{\Omega_j} + \sum_k \langle \lambda_{w,jk}^n, \mathbf{v}_1 \cdot \nu \rangle_{\Gamma_{jk}} = 0, \tag{3.3b}$$

$$-\frac{(\Phi S_j^n, w_2)_{\Omega_j} - (\Phi S_j^{n-1}, w_2)_{\Omega_j}}{\Delta t^n} - (\nabla \cdot \mathbf{Q}_{o,j}^n, w_2)_{\Omega_j} \tag{3.3c}$$
$$= (q_{\text{ext},o}^n, w_2)_{\Omega_j} + (q_{m,o}^n, w_2)_{\Omega_j},$$

$$\left(\frac{\mathbf{Q}_{o,j}^n}{\tilde{\Lambda}_o(S_j^n)}, \mathbf{v}_2\right)_{\Omega_j} - (\Psi_{o,j}^n, \operatorname{div} \mathbf{v}_2)_{\Omega_j} + \sum_k \langle \lambda_{o,jk}^n, \mathbf{v}_2 \cdot \nu \rangle_{\Gamma_{jk}} = 0, \tag{3.3d}$$

$$(P_c(S_j^n), w_3)_{\Omega_j} = (\Psi_{c,j}^n, w_3)_{\Omega_j} + ((\rho_o - \rho_w) g z, w_3)_{\Omega_j}, \tag{3.3e}$$

where $\mathbf{v}_1, \mathbf{v}_2 \in \mathbf{V}_j$ and $w_1, w_2, w_3 \in \mathbf{W}_j$.

3.3 The Iterative Method

In order to define an iterative method for solving the above system [9], [8] it is convenient to replace (3.3b) and (3.3d) by the Robin transmission boundary condition

$$-\beta \mathbf{Q}_{\alpha,j} \cdot \nu_j + \Psi_{\alpha,j} = \beta \mathbf{Q}_{\alpha,k} \cdot \nu_k + \Psi_{\alpha,k}, \qquad x \in \Gamma_{jk} \subset \partial\Omega_j,\ \alpha = w, o, \tag{3.4a}$$
$$-\beta \mathbf{Q}_{\alpha,k} \cdot \nu_k + \Psi_{\alpha,k} = \beta \mathbf{Q}_{\alpha,j} \cdot \nu_j + \Psi_{\alpha,j}, \qquad x \in \Gamma_{kj} \subset \partial\Omega_k,\ \alpha = w, o, \tag{3.4b}$$

where β is a positive (normally chosen to be a constant) function on $\cup\Gamma_{jk}$.

Now we formulate an iterative version of the finite element approximation of (3.3) with consistency conditions given by (3.4). Consider the Lagrange multiplier to be $\lambda_{\alpha,jk}$, $\alpha = w, o$ as seen from Ω_j and $\lambda_{\alpha,kj}$, $\alpha = w, o$ as seen from Ω_k. Then, modify (3.4) to read

$$-\beta \mathbf{Q}_{\alpha,j} \cdot \nu_j + \lambda_{\alpha,jk} = \beta \mathbf{Q}_{\alpha,k} \cdot \nu_k + \lambda_{\alpha,kj}, \qquad x \in \Gamma_{jk} \subset \partial\Omega_j,\ \alpha = w, o,$$
$$-\beta \mathbf{Q}_{\alpha,k} \cdot \nu_k + \lambda_{\alpha,kj} = \beta \mathbf{Q}_{\alpha,j} \cdot \nu_j + \lambda_{\alpha,jk}, \qquad x \in \Gamma_{kj} \subset \partial\Omega_k,\ \alpha = w, o,$$

so that

$$\langle \lambda_{\alpha,jk}, \mathbf{v} \cdot \nu_j \rangle_{\Gamma_{jk}} = \langle \beta(\mathbf{Q}_{\alpha,j} \cdot \nu_j + \mathbf{Q}_{\alpha,k} \cdot \nu_k) + \lambda_{\alpha,kj}, \mathbf{v} \cdot \nu_j \rangle_{\Gamma_{jk}}, \quad \alpha = w, o.$$

The objective of a domain decomposition iterative method is to localize the calculations to problems over smaller domains than Ω. Here, it is feasible to localize to each Ω_j by evaluating the quantities in (3.3) related to Ω_j at the new iterate level and those in (3.3) related to neighboring subdomains Ω_k such that $\Gamma_{jk} \neq \emptyset$ at the previous iteration level. Specifically, the algorithm would be as follows: let, for all j and k,

$$S_j^{n-1} \in \mathbf{W}_j, \mathbf{Q}_{\alpha,j}^{n-1} \in \mathbf{V}_j, \quad \Psi_{\alpha,j}^{n-1} \in \mathbf{W}_j, \quad \lambda_{\alpha,jk}^{n-1} \in \Lambda_{\alpha,jk}, \quad \lambda_{\alpha,kj}^{n-1} \in \Lambda_{\alpha,kj}, \alpha = w, o,$$

($\lambda^0_{jk} = \lambda^0_{kj}$ seems natural) be the solution of the discretized system of equations at some discrete time (we introduce in the notation the superscript i which is an iteration counter).

Then the solution propagated by one time step is given as the limit as $i \to \infty$ of recursive solutions of the equations

$$\frac{(\Phi S^{n,i}_j, w_1)_{\Omega_j} - (\Phi S^{n-1}_j, w_1)_{\Omega_j}}{\Delta t^n} - (\nabla \cdot \mathbf{Q}^{n,i}_{w,j}, w_1)_{\Omega_j} \tag{3.5a}$$
$$= (q^{n,i}_{\mathrm{ext},w}, w_1)_{\Omega_j} + (q^{n,i}_{m,w}, w_1)_{\Omega_j},$$

$$\left(\frac{\mathbf{Q}^{n,i}_{w,j}}{\tilde{\Lambda}_w(S^{n,i-1}_j)}, \mathbf{v}_1\right)_{\Omega_j} - (\Psi^{n,i}_{w,j}, \operatorname{div} \mathbf{v}_1)_{\Omega_j} + \sum_k \langle \beta \mathbf{Q}^{n,i}_{w,j} \cdot \nu_j, \mathbf{v}_1 \cdot \nu_j \rangle_{\Gamma_{jk}} \tag{3.5b}$$
$$= -\sum_k \langle \beta \mathbf{Q}^{n,i-1}_{w,k} \cdot \nu_k + \lambda^{n,i-1}_{w,kj}, \mathbf{v}_1 \cdot \nu_j \rangle_{\Gamma_{jk}},$$

$$\frac{(\Phi S^{n,i}_j, w_2)_{\Omega_j} - (\Phi S^{n-1}_j, w_2)_{\Omega_j}}{\Delta t^n} - (\nabla \cdot \mathbf{Q}^{n,i}_{o,j}, w_2)_{\Omega_j} \tag{3.5c}$$
$$= (q^{n,i}_{\mathrm{ext},o}, w_2)_{\Omega_j} + (q^{n,i}_{m,o}, w_2)_{\Omega_j},$$

$$\left(\frac{\mathbf{Q}^{n,i}_{o,j}}{\tilde{\Lambda}_o(S^{n,i-1}_j)}, \mathbf{v}_2\right)_{\Omega_j} - (\Psi^{n,i}_{o,j}, \operatorname{div} \mathbf{v}_2)_{\Omega_j} + \sum_k \langle \beta \mathbf{Q}^{n,i}_{o,j} \cdot \nu_j, \mathbf{v}_2 \cdot \nu_j \rangle_{\Gamma_{jk}} \tag{3.5d}$$
$$= -\sum_k \langle \beta \mathbf{Q}^{n,i-1}_{o,k} \cdot \nu_k + \lambda^{n,i-1}_{o,kj}, \mathbf{v}_2 \cdot \nu_j \rangle_{\Gamma_{jk}}.$$

The Lagrange multipliers are updated according to

$$\lambda^{n,i}_{w,jk} = \beta(\mathbf{Q}^{n,i}_{w,j} \cdot \nu_j + \mathbf{Q}^{n,i-1}_{w,k} \cdot \nu_k) + \lambda^{n,i-1}_{w,kj}, \tag{3.5e}$$
$$\lambda^{n,i}_{o,jk} = \beta(\mathbf{Q}^{n,i}_{o,j} \cdot \nu_j + \mathbf{Q}^{n,i-1}_{o,k} \cdot \nu_k) + \lambda^{n,i-1}_{o,kj}, \tag{3.5f}$$

and finally the equation for the capillary pressure is linearized:

$$\left(\frac{\partial P_c(S^{n,i-1}_j)}{\partial S}(S^{n,i}_j - S^{n,i-1}_j), w_3\right)_{\Omega_j} \tag{3.5g}$$
$$= (\Psi^{n,i}_{o,j} - \Psi^{n,i}_{w,j}, w_3)_{\Omega_j} + ((\rho_o - \rho_w)gz, w_3)_{\Omega_j} - (P_c(S^{n,i-1}_j), w_3)_{\Omega_j}.$$

We still have to explain how the matrix source terms are incorporated into the iterative procedure. We will postpone this discussion to §4.

We have been able to prove the following theorem concerning the convergence of the iterative procedure defined above in the simplified context where matrix blocks are suppressed from the model.

Theorem 3.1 (Convergence of the Iterative Procedure) *Suppose that a smooth solution of the system (2.1) exists. Then, there exists a constant t^* such that, when $\Delta t \le t^*$,*

1) the iterative scheme (3.5) converges; i.e., there exists

$$\{S^n_j \in \mathbf{W}_j, \mathbf{Q}^n_{\alpha,j} \in \mathbf{V}_j, \Psi^n_{\alpha,j} \in \mathbf{W}_j, \lambda^n_{\alpha,jk} \in \Lambda_{\alpha,jk}\}$$

such that

$$\|S_j^{n,i} - S_j^n\| + \|\mathbf{Q}_{\alpha,j}^{n,i} - \mathbf{Q}_{\alpha,j}^n\| + \|\Psi_{\alpha,j}^{n,i} - \Psi_{\alpha,j}^n\| + \|\lambda_{\alpha,jk}^{n,i} - \lambda_{\alpha,jk}^n\| \to 0$$

as $i \to \infty$*; moreover,* $\lambda_{\alpha,jk}^n = \lambda_{\alpha,kj}^n$, $\alpha = w, o$*; and*

2) *the above limit converges to the smooth solution in the sense there is a constant* c *such that*

$$\|S(t^n) - S^n\| + \|\mathbf{Q}_\alpha(t^n) - \mathbf{Q}_\alpha^n\| + \|\Psi_\alpha(t^n) - \Psi_\alpha^n\| \le c(\Delta t + h),$$

where $\Theta^n(x) = \Theta_j^n(x)$*, for* $x \in \Omega_j$ *and* h *represents the partition of* Ω*.*

A rigorous proof of the above statement will appear elsewhere.

4 THE COMPUTATIONAL ALGORITHM

Our numerical procedure will combine a computationally inexpensive finite difference procedure to solve the local problems associated with each matrix block with a hybridized mixed finite element method applied to the global fracture system problem. The fracture and matrix systems cannot be handled sequentially, since a small change in the boundary values on each matrix block can cause flow of a volume of fluid that is large in comparison to the volume of the fractures. The matrix-fracture interaction for the medium block model can be handled implicitly by a linearization of the matrix problems to be made precise below. The final procedure requires solution of many small linear systems, each corresponding to an element of the discretized fracture system and its associated matrix block. The solution of these small and uncoupled linear systems can be handled easily by a parallel machine.

Discretize the space variables by defining grids over Ω and over each matrix block Ω_x. We consider Ω and Ω_x, $x \in \Omega$, to be rectangular parallelepipeds; more general domains can be treated by either finite difference or finite element techniques quite analogous to the methods to be described herein. Suppose that $\Omega = [0, D_1] \times [0, D_2] \times [0, D_3]$. Then, divide each D_j into N_j intervals, which for simplicity we take to be of equal size $H_j = D_j/N_j$, $j = 1, 2, 3$. Thus, the set

$$\mathcal{G}_f = \{x_L = (L_1H_1{+}H_1/2, L_2H_2{+}H_2/2, L_3H_3{+}H_3/2) : L_j = 0, 1, \ldots, N_j{-}1; j = 1, 2, 3\},$$

consists of the centers of the elements of the mixed method.

Again for notational convenience assume that the matrix blocks are all of the same size and consider a grid defined on the matrix block Ω_x which will be used in a finite difference discretization of the matrix equations to be discussed below. Let h_j and n_j be analogous to H_j and N_j and set

$$\mathcal{G}_m = \{y_\ell = (\ell_1h_1, \ell_2h_2, \ell_3h_3) : \ell_j = 0, 1, 2, \ldots, n_j,\ j = 1, 2, 3\}.$$

Also, let

$$\mathcal{I}_m = \{y_\ell = (\ell_1h_1, \ell_2h_2, \ell_3h_3) : \ell_j = 1, 2, \ldots, n_j - 1,\ j = 1, 2, 3\}$$

indicate the interior nodes and $\partial\mathcal{G}_m = \mathcal{G}_m \setminus \mathcal{I}_m$ the boundary nodes. (Advantage should be taken of any symmetry of the solution on a matrix block to allow the solution to be computed only at necessary nodes.)

An approximation to a function Θ related to the the fracture system at a point $x_L \in \mathcal{G}_f$ will be denoted by

$$\Theta_L^n \approx \Theta(x_L, t^n).$$

and for a function θ associated with the block at the point $x_L \in \mathcal{G}_f$, denote the approximation to θ at $y_\ell \in \mathcal{G}_m$ by

$$\theta_{L,\ell}^n \approx \theta(x_{L,\ell}, t^n),$$

where $x_{L,\ell} = x_L - y_\ell$ (this places a top corner of the block at x_L).

The matrix equations will be completely linearized, but *not* the fracture equations. The discrete matrix system is directly solvable. The four parts of the algorithm below uncouple the calculations related to the matrix blocks from those of the fracture calculation:

i) Initialization. For each L and ℓ, set

$$\Psi_{c,L}^0 = \Psi_{\text{init},c}(x_L), \qquad S_L^0 = P_c^{-1}\left(\Psi_{c,L}^0 + (\rho_o - \rho_w)g\, z(x_L)\right),$$
$$\psi_{c,L,\ell}^0 = \psi_{\text{init},c}(x_{L,\ell}), \qquad s_{L,\ell}^0 = p_c^{-1}\left(\psi_{c,L,\ell}^0 + (\rho_o - \rho_w)g\, z(x_{L,\ell})\right).$$

ii) Matrix system. For each L, ℓ, and for $n \geq 1$, find $\{\bar{\psi}_{c,L,\ell}^n, \bar{\psi}_{w,L,\ell}^n\}$ by solving

$$\phi(x_{L,\ell})\,\frac{\bar{\psi}_{c,L,\ell}^n - \psi_{c,L,\ell}^{n-1}}{p_c'(s^{n-1})\Delta t^n} - \nabla_{h,L,\ell}\cdot\left[\tilde{\lambda}_w(s^{n-1})\nabla_{h,L,\ell}\bar{\psi}_w^n\right] = 0 \quad \text{if } y_\ell \in \mathcal{I}_m \tag{4.1a}$$
$$-\nabla_{h,L,\ell}\cdot\left[\tilde{\lambda}(s^{n-1})\nabla_{h,L,\ell}\bar{\psi}_w^n + \tilde{\lambda}_o(s^{n-1})\nabla_{h,L,\ell}\bar{\psi}_c^n\right] = 0 \quad \text{if } y_\ell \in \mathcal{I}_m, \tag{4.1b}$$
$$\bar{\psi}_{c,L,\ell}^n = \Psi_{c,L}^{n-1} \quad \text{and} \quad \bar{\psi}_{w,L,\ell}^n = \Psi_{w,L}^{n-1} \quad \text{if } y_\ell \in \partial\mathcal{G}_m, \tag{4.1c}$$

and determine $\{\breve{\psi}_{c,L,\ell}^n, \breve{\psi}_{w,L,\ell}^n\}$ and $\{\hat{\psi}_{c,L,\ell}^n, \hat{\psi}_{w,L,\ell}^n\}$ by solving

$$\phi(x_{L,\ell})\,\frac{\breve{\psi}_{c,L,\ell}^n}{p_c'(s^{n-1})\Delta t^n} - \nabla_{h,L,\ell}\cdot\left[\tilde{\lambda}_w(s^{n-1})\nabla_{h,L,\ell}\breve{\psi}_w^n\right] = 0 \quad \text{if } y_\ell \in \mathcal{I}_m, \tag{4.2a}$$
$$-\nabla_{h,L,\ell}\cdot\left[\tilde{\lambda}(s^{n-1})\nabla_{h,L,\ell}\breve{\psi}_w^n + \tilde{\lambda}_o(s^{n-1})\nabla_{h,L,\ell}\breve{\psi}_c^n\right] = 0 \quad \text{if } y_\ell \in \mathcal{I}_m, \tag{4.2b}$$
$$\breve{\psi}_{c,L,\ell}^n = 1 \quad \text{and} \quad \breve{\psi}_{w,L,\ell}^n = 0 \quad \text{if } y_\ell \in \partial\mathcal{G}_m, \tag{4.2c}$$

and

$$\phi(x_{L,\ell})\,\frac{\hat{\psi}_{c,L,\ell}^n}{p_c'(s^{n-1})\Delta t^n} - \nabla_{h,L,\ell}\cdot\left[\tilde{\lambda}_w(s^{n-1})\nabla_{h,L,\ell}\hat{\psi}_w^n\right] = 0 \quad \text{if } y_\ell \in \mathcal{I}_m,$$
$$-\nabla_{h,L,\ell}\cdot\left[\tilde{\lambda}(s^{n-1})\nabla_{h,L,\ell}\hat{\psi}_w^n + \tilde{\lambda}_o(s^{n-1})\nabla_{h,L,\ell}\hat{\psi}_c^n\right] = 0 \quad \text{if } y_\ell \in \mathcal{I}_m,$$
$$\hat{\psi}_{c,L,\ell}^n = 0 \quad \text{and} \quad \hat{\psi}_{w,L,\ell}^n = 1 \quad \text{if } y_\ell \in \partial\mathcal{G}_m, \tag{4.3a}$$

where

$$\nabla_{h,L,\ell}\cdot\left[\tilde\lambda_\alpha(s^{n-1})\nabla_{h,L,\ell}\psi^n\right]=\sum_{j=1}^{3}\frac{1}{h_j^2}\left\{\tilde\lambda_\alpha\left(\frac{s^{n-1}_{L,\ell+e_j}+s^{n-1}_{L,\ell}}{2}\right)(\psi^n_{L,\ell+e_j}-\psi^n_{L,\ell})\right.$$
$$\left.-\tilde\lambda_\alpha\left(\frac{s^{n-1}_{L,\ell}+s^{n-1}_{L,\ell-e_j}}{2}\right)(\psi^n_{L,\ell}-\psi^n_{L,\ell-e_j})\right\}.$$

These equations are linear, since the mobilities and p_c' are evaluated at the previous time level. The matrix potentials $\psi^n_{c,L,\ell}$ and $\psi^n_{w,L,\ell}$ are defined below in (4.7) and (4.8); they satisfy the expected equations, namely (4.10). Equations (4.1 define a particular solution to the linear equations, while (4) and (4.3) give solutions to the homogeneous problems which describe unit changes in the boundary conditions.

iii) The Iterative Procedure.

a) The matrix source term. For each L and $n\ge 1$, compute

$$\tilde\psi^{n,i}_{c,L,\ell}=\bar\psi^n_{c,L,\ell}+\left(\Psi^{n,i}_{c,L}-\Psi^{n-1}_{c,L}\right)\breve\psi^n_{c,L,\ell}+\left(\Psi^{n,i}_{w,L}-\Psi^{n-1}_{w,L}\right)\hat\psi^n_{c,L,\ell},$$
$$\tilde s^{n,i}_{L,\ell}=p_c^{-1}\left(\tilde\psi^{n,i}_{c,L,\ell}+(\rho_o-\rho_w)gz(x_{L,\ell})\right),$$
$$q^{n,i}_{m,w,L}=-\frac{1}{|\Omega_x|}\sum_\ell\phi(x_{L,\ell})\frac{\tilde s^{n,i}_{L,\ell}-s^{n-1}_{L,\ell}}{\Delta t^n}V_\ell,\tag{4.4}$$

where V_ℓ is the volume element associated with the grid point ℓ. The quantity $q^n_{m,w,L}$ is given implicitly in terms of the fracture potentials at the nth time level; however, in view of (4.7) and (4.9) below, (4.4) is clearly a discretization of (2.4).

b) Fracture System. For each L and $n\ge 1$, solve the nonlinear system of equations (3.5) using the iterative method described in §3 for S^n_L, $\mathbf{Q}^n_{\alpha,L}$, $\Psi^n_{\alpha,L}$, λ^n_α, $\alpha=w,o$ by computing $q^{n,i}_{m,w,L}$ employing iii) above (for simplicity, we denote the set of four Lagrange multipliers associated to each element by λ_α). The no–flow boundary conditions of (3.2b) and (3.2d) are imposed by considering virtual elements outside the computational region such that

$$\lambda^{n,i}_\alpha=\lambda^{n,i}_{\alpha,L\mp e_j},\qquad \alpha=w,o,\tag{4.5}$$
$$\mathbf{Q}^n_{\alpha,L\pm e_j}=0,\qquad \alpha=w,o,\tag{4.6}$$

if $x_{L\pm e_j}$ is outside the reservoir.

Since physically meaningful capillary pressures are nonnegative, the capillary functions should be extended vertically downward at $S_{\max}$ or $s_{\max}$.

iv) Matrix update. For each L,ℓ, and $n\ge 1$, let

$$\psi^n_{c,L,\ell}=\bar\psi^n_{c,L,\ell}+\left(\Psi^n_{c,L}-\Psi^{n-1}_{c,L}\right)\breve\psi^n_{c,L}+\left(\Psi^n_{w,L}-\Psi^{n-1}_{w,L}\right)\hat\psi^n_{c,L,\ell},\tag{4.7}$$
$$\psi^n_{w,L,\ell}=\bar\psi^n_{w,L,\ell}+\left(\Psi^n_{c,L}-\Psi^{n-1}_{c,L}\right)\breve\psi^n_{w,L}+\left(\Psi^n_{w,L}-\Psi^{n-1}_{w,L}\right)\hat\psi^n_{w,L,\ell},\tag{4.8}$$
$$s^n_{L,\ell}=p_c^{-1}\left(\psi^n_{c,L,\ell}+(\rho_o-\rho_w)gz(x_{L,\ell})\right).\tag{4.9}$$

This completes the time step.

The above algorithm can be implemented sequentially. The following discrete matrix problem has been solved:

$$\phi(x_{L,\ell})\frac{\psi^n_{c,L,\ell}-\psi^{n-1}_{c,L,\ell}}{p'_c(s^{n-1})\Delta t^n}-\nabla_{h,L,\ell}\cdot\left[\tilde{\lambda}_w(s^{n-1})\nabla_{h,L,\ell}\psi^n_w\right]=0 \quad \text{if } y_\ell\in\mathcal{I}_m, \tag{4.10a}$$

$$-\nabla_{h,L,\ell}\cdot\left[\tilde{\lambda}(s^{n-1})\nabla_{h,L,\ell}\psi^n_w+\tilde{\lambda}_o(s^{n-1})\nabla_{h,L,\ell}\psi^n_c\right]=0 \quad \text{if } y_\ell\in\mathcal{I}_m, \tag{4.10b}$$

$$\psi^n_{c,L,\ell}=\Psi^n_{c,L} \quad \text{and} \quad \psi^n_{w,L,\ell}=\Psi^n_{w,L} \quad \text{if } y_\ell\in\partial\mathcal{G}_m. \tag{4.10c}$$

Assuming that the wetting fluid is the denser, it should be noted that the block associated with the fracture point x_L is interpreted to lie below x_L for imbibition, the case we have treated. For drainage it should be placed above x_L; otherwise, fluid is trapped by the numerical simulation as P_c tends to zero.

The numerical convergence of the iterative method just described is measured in terms of a relative error defined in terms of the ℓ_2 norm of variables describing the flow as the number of iterations is successively doubled.

5 PARALLEL IMPLEMENTATIONS

We developed a serial code based on the algorithm described in the previous section and validated it against another code developed independently (which uses finite differences) in [16]. In order to minimize execution time we use two techniques to reduce the number of iterations required for convergence. Different procedures are required depending on the time step number. After two time steps have been solved, a quadratic extrapolation in time [14], [13] reduces drastically the number of iterations required for convergence. A different procedure has to be adopted for the two initial time steps. We used a variant of the above method. Instead of extrapolating in time, we consider a spatial hierarchical extrapolation. We solve the problem in a family of nested grids, interpolate the solution according to the finite element method in use on each grid, and then we use a quadratic extrapolation as a function of the grid size. Figure 1 illustrates this procedure. Using the hierarchical extrapolation we typically reduce by a factor of two the execution time for the initial time steps.

We consider speedup studies through simulations of waterflooding calculations in a "five-spot" geometry with gravity effects neglected.

For computational simplicity, the fracture calculations are two-dimensional over Ω, though the matrix calculations would remain three-dimensional over each Ω_x if gravity were not ignored. Initially, the reservoir contains 75% oil and 25% water. Water is injected uniformly into the reservoir along one corner at a rate of one pore-volume every five years.

The following data are held fixed for the computational results exhibited below:

Fluid properties		
Viscosity	$\mu_w = .5$ cP	$\mu_o = 2$ cP
Density	$\rho_w = 1$ g/ cm ^{3}	$\rho_o = .7$ g/ cm ^{3}
Absolute Permeabilities	$K = 1$ darcy	$k = 0.05$ darcy
Porosities	$\Phi = .01$	$\phi = .1$
Residual Saturations (matrix)	$s_{ro} = .15$	$s_{rw} = .2$
Residual Saturations (fractures)	$S_{ro} = 0$	$S_{rw} = 0$

The capillary pressure functions were assumed in the form

$$\begin{aligned} P_c(S) &= (1-S)\{\gamma(S^{-1}-1)+\Theta\}, \\ p_c(s) &= \alpha((s-s_{rw})^{-2}-\beta(1-s)^{-2}), \\ s_0 &= 1-s_{ro}, \qquad \beta = s_{ro}^2(s_0-s_{rw})^{-2}, \\ \gamma &= 2.0\times 10^4 \,\text{dynes}/\text{cm}^2, \quad \Theta = 100\,\text{dynes}/\text{cm}^2, \\ \alpha &= 3.0\times 10^3\,\text{dynes}/\text{cm}^2. \end{aligned}$$

The relative permeability functions in the fractures were chosen to be linear, with the residual saturations taken to be zero:

$$K_{ro}(S) = 1-S, \qquad K_{rw}(S) = S.$$

In the matrix blocks the relative permeabilities functions were taken to be

$$\begin{aligned} k_{ro}(s) &= \{1-(1-s_{ro})^{-1}s\}^2, \\ k_{rw}(s) &= (1-s_{rw})^{-2}(s-s_{rw})^2. \end{aligned}$$

5.1 MIMD Implementation

The MIMD version of the serial code described in the previous sections is implemented through CMMD, a library of the CM-5 and uses SPARC microprocessors. See [27] for additional information about a CM-5 system. A hostless programming model is used, in which each node receives the same copy of a code. The computer code is written in the *C* language and the driver of the program is written in terms of function pointers. This allows us to assign different functions to distinct subdomains, such that subdomain dependent procedures (like injection of fluid in specified positions and imposition of boundary conditions) can be handled.

The computational domain is decomposed into rectangular regions. Each of the subdomains (which in general will contain several elements of the discretized fracture system of equations) is assigned to a different processor. Each processor allocates memory for the elements contained in its subdomain and for a buffer zone consisting of one layer of elements outside the subdomain. The elements contained in each rectangular region are processed sequentially, using the algorithm described in §3 with a modification which allows exchange of information between nearest neighbor subdomains. Once one step of the iterative procedure is performed on each element within a rectangular

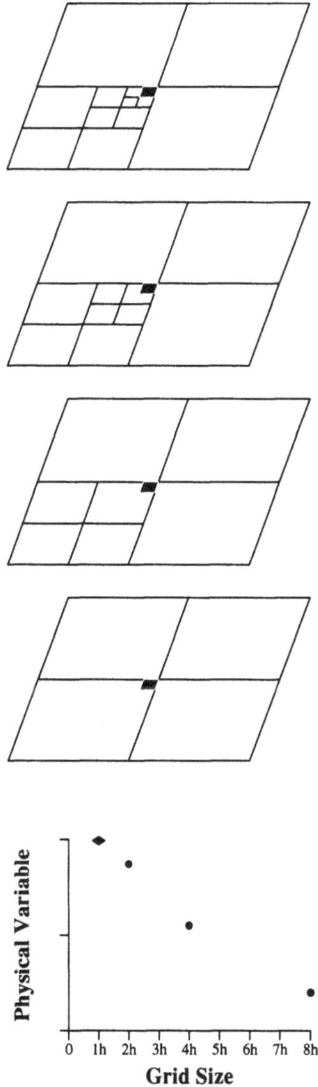

Figure 1: The hierarchical extrapolation. A guess for the iterative method is computed using an extrapolation of solutions of a given problem in coarser grids as a function of the mesh size. Given a problem on a grid with mesh size h the guess is computed using approximate solutions for the the same problem on grids with sizes $2h$, $4h$, and $8h$.

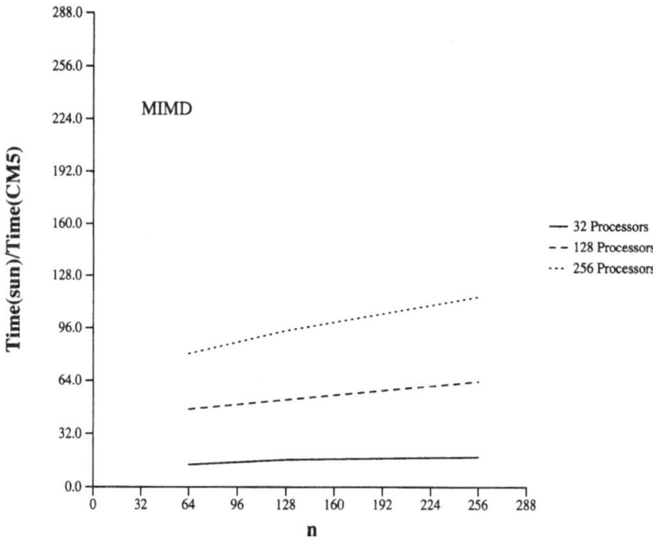

Figure 2: The ratio of the time spent by a SUN SPARC station to finish a simulation divided by the time spent by a partition of the CM-5 (running the MIMD version of our code) with variable number of processors is plotted against n, the number of elements in one direction of the grid. The physical size of the reservoir is increased with n, keeping the mesh size fixed. For the largest problem considered the speedup obtained is about half of the number of processors used.

region then, through a sequence of grid shifts (right, left, up and down), data on the boundary of subdomains is sent (received) to (from) neighboring subdomains. The boundary conditions (4.5) and (4.6) are also set at this stage of the computation.

We addressed the problem of the speedup obtained with the CM-5 in two studies. First, we compared execution times for simulations performed in the CM-5 with the same simulations performed in a SUN SPARC station. In Figure 2 we plot the ratio of the time spent by a SUN SPARC station to the time spent by different partitions of the CM-5 as a function of the problem size (represented in the plot by the number n of points in one direction of the grid), with mesh size kept fixed. Note in Figure 2 that for the largest grid considered (256×256) the speedup obtained is about half of the number of processors used. Next, we considered the speedup curve for simulations with large grids. We considered the ratio of the time spent by partitions of the CM-5 with 256 and 512 processors (to perform a family of simulations with increasing physical size) to the time spent by 128 processors as a function of the number of processors used. The result of this study is reported in Figure 4. Note in Figure 4 that, as the problem size is increased, the closer the speedup curve gets to the perfect (linear) speedup.

5.2 SIMD Implementation

SIMD control in a CM-5 system is achieved through the notion of virtual processors and implemented with data-parallel programing languages. Our code was developed

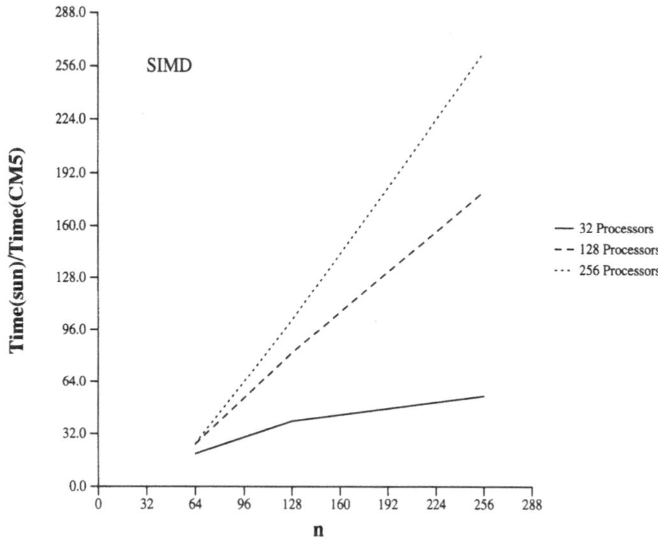

Figure 3: The ratio of the time spent by a SUN SPARC station to finish a simulation divided by the time spent by a partition of the CM-5 (running the SIMD version of our code) with variable number of processors is plotted against n, the number of elements in one direction of the grid. For problems with small grids increasing the number of processors has little effect on execution time. For the largest problem considered the speedup obtained is more than the number of processors used.

using the language C^*. Vector processors were used to run the SIMD version of our code.

The C^* program is quite similar to serial code. We used grid communication within C^* to perform the necesssary exchange of information and the "where" statement to set boundary conditions.

Again, we considered the problem of the speedup obtained with the CM-5 in two studies. An study analogous to the one described in Figure 2 for the MIMD version of our code is the content of Figure 3. Note in Figure 3 that as n is increased the performance of the CM-5 increases. The speedup obtained for the largest problem size considered (256×256) in this study is more than the number of vector processors used. Next, we considered the speedup curve for simulations with large grids. This study appears in Figure 4. As explained above, the ratio of the time spent by partitions of the CM-5 with 256 and 512 processors to the time spent by 128 processors is plotted against the number of processors used. As we noted above for the MIMD version of our code, as the problem size is increased the closer the speedup curve gets to the perfect (linear) speedup.

Figure 4 also allows us to compare the speedup obtained with the two parallel implementations reported here. Although the execution times of the SIMD version are about half of the MIMD version (due to the use of vector processing units) Figure 4 shows a better scalability of the MIMD version.

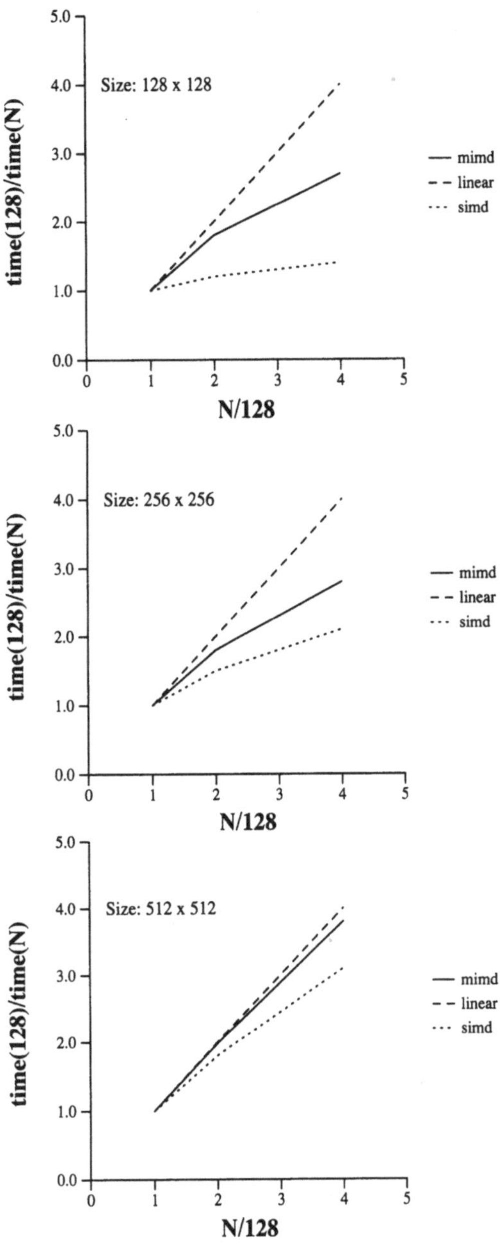

Figure 4: Speedup curves obtained with the MIMD and SIMD versions of the new numerical scheme. As the problem size is increased both versions display a better performance.

6 CONCLUSIONS AND OPEN PROBLEMS

We proposed a new numerical method to solve the system of equations governing immiscible flow in naturally fractured reservoirs in massively parallel computers. The numerical method uses spatial decomposition in the context of an iterative procedure to solve the global problem associated with the part of the system governing fracture flow. By decomposing the domain into the elements of the mixed finite element method used in the discretization of fracture equations the numerical scheme can be easily translated into a computer code written in data-parallel languages.

The new numerical procedure was implemented first in a serial machine, and numerical simulations were performed to validate the code. Then, the serial code was restructured and implemented in a CM-5 system, both in MIMD and SIMD modes. We found a good scalability of the two versions of the parallel code for problems with grid sizes relevant in applications.

We established the convergence of the new numerical procedure for a simplified version of the model discussed here, and the proof of convergence for the full system constitutes an interesting open problem. From the numerical point of view, the new numerical method will allow us to study multi-length scale, stochastic, double porosity models, due to the high resolution provided by the CM-5 through the use of fine computational grids. Obviously, full three-dimensional fluid flow simulations remain as one of the most interesting challenges of our research area; such simulations are currently being pursued by the authors.

REFERENCES

[1] Arnold D. N., Brezzi F. Mixed and nonconforming finite element methods: implementation, postprocessing and error estimates. *RAIRO*, 19:7–32, 1985.

[2] Brezzi F., Douglas J. Jr., Durán R., Fortin M. Mixed finite elements for second order elliptic problems in three variables. *Numer. Math.*, 51:237–250, 1987.

[3] Brezzi F., Douglas J. Jr., Fortin M., Marini L. D. Efficient rectangular mixed finite elements in two and three space variables. *R.A.I.R.O. Modélisation Mathématique et Analyse Numérique*, 21:581–604, 1987.

[4] Brezzi F., Douglas J. Jr., Marini L. D. Two families of mixed finite elements for second order elliptic problems. *Numerische Mathematik*, 47:217–235, 1985.

[5] Brezzi F., Douglas J. Jr., Marini L. D. Variable degree mixed methods for second order elliptic problems. *Matemática Aplicada e Computacional*, 4:19–34, 1985.

[6] Chen Z., Douglas J. Jr. Prismatic mixed finite elements for second order elliptic problems. *Calcolo*, 26:135–148, 1989.

[7] Cowsar L. C., Wheeler M. F. Parallel domain decomposition method for mixed finite elements for elliptic partial differential equations. In *Proceedings of the Fourth International Symposium on Domain Decomposition Methods for Partial Differential Equations*, SIAM, Philadelphia, 1991. R. Glowinski, Y. Kuznetsov, G. Meurant, J. Périaux, and O. B. Widlund, eds.

[8] Després B. *Domain decomposition method and the Helmholz problem*, pages 44–52. SIAM, Philadelphia, 1991. G. Cohen, L. Halpern, and P. Joly (eds.).

[9] Després B. *Méthodes de décomposition de domaines pour les problèmes de propagation d'ondes en régime harmonique.* PhD thesis, Université Paris IX Dauphine, UER Mathématiques de la Décision, 1991.

[10] Després B., Joly P., Roberts J. E. *A domain decomposition method for the harmonic Maxwell equations*, pages 475–484. Elsevier Science Publishers B. V. (North-Holland), Amsterdam, 1992. R. Beauwens and P. de Groen (eds.).

[11] Douglas J., Jr., Arbogast T. Dual porosity models for flow in naturally fractured reservoirs. In *Dynamics of Fluids in Hierarchical Porous Formations*, pages 177–221, Academic Press, London, 1990. J. H. Cushman, ed.

[12] Douglas J. Jr., Arbogast T., Paes Leme P. J., Hensley J. L., Nunes N. P. Immiscible displacement in vertically fractured reservoirs. *Transport in Porous Media.* To appear, 1993.

[13] Douglas J. Jr., Dupont T., Ewing R. E. Incomplete iteration for time–stepping a nonlinear parabolic Galerkin method. *SIAM J. Numer. Anal.*, 16:503–522, 1979.

[14] Douglas J. Jr., Dupont T., Percell P. A time–stepping method for Galerkin approximations for nonlinear parabolic equations. In *Numerical Analysis, Dundee 1977*, Springer-Verlag, Berlin, 1978.

[15] Douglas J. Jr., Hensley J. H., Paes Leme P. J. A study of the effect of inhomogeneities on immiscible flow in naturally fractured reservoirs. In *Porous Media*, Birkhäuser, Basel, 1993.

[16] Douglas J., Jr., Hensley J. L., Arbogast T. A dual–porosity model for waterflooding in naturally fractured reservoirs. *Computer Methods in Applied Mechanics and Engineering*, 87:157–174, 1991.

[17] Douglas J. Jr., Paes Leme P. J., Roberts J. E., Wang J. A parallel iterative procedure applicable to the approximate solution of second order partial differential equations by mixed finite element methods. *Numerische Mathematik.* To appear, 1993.

[18] Ewing R. E., Wang J. Analysis of multilevel decomposition iterative methods for mixed finite element methods. *R.A.I.R.O. Modélisation Mathématique et Analyse Numérique.* Submitted.

[19] Ewing R. E., Wang J. Analysis of the Schwarz algorithm for mixed finite element methods. *R.A.I.R.O. Modélisation Mathématique et Analyse Numérique*, 26:739–756, 1992.

[20] Fraeijs de Veubeke B. X. Displacement and equilibrium models in the finite element method. In *Stress Analysis*, John Wiley, New York, 1965. O. C. Zienkiewicz and G. Holister (eds.).

[21] Fraeijs de Veubeke B. X. Stress function approach. International Congress on the Finite Element Method in Structural Mechanics, Bournemouth, 1975.

[22] Glimm J., Lindquist B., Pereira F., Peierls R. The fractal hypothesis and anomalous diffusion. *Matemática Aplicada e Computacional*, 11:189–207, 1992.

[23] Glimm, J., Lindquist B., Pereira F., Zhang Q. A theory of macrodispersion for the scale up problem. *Advances in Water Resources*. To appear.

[24] Glowinski R., Kinton W., Wheeler M. F. Acceleration of domain decomposition algorithms for mixed finite elements by multi-level methods. In *Third International Symposium on Domain Decomposition Methods for Partial Differential Equations*, pages 263–290, SIAM, Philadelphia, 1990. R. Glowinski (ed.).

[25] Glowinski R., Wheeler M. F. *Domain decomposition and mixed finite element methods for elliptic problems*, pages 144–172. SIAM, Philadelphia, 1988. R. Glowinski, G. Golub, G. Meurant, and J. Periaux (eds.).

[26] Nedelec J. C. Mixed finite elements in $\mathbf{R}^3$. *Numer. Math.*, 35:315–341, 1980.

[27] Palmer J., Steele G. L. Jr. Connection Machine Model CM-5 system overview. In *Proceedings of the Fourth Symposium on the Frontiers of Massively Parallel Computation*, pages 474–483, IEEE Computer Society Press, Los Alamitos, 1992.

[28] Raviart P. A., Thomas J. M. A mixed finite element method for second order elliptic problems. In *Mathematical Aspects of the Finite Element Method*, pages 292–315, Springer-Verlag, Berlin and New York, 1977. I. Galligani and E. Magenes, eds.

[29] Thomas J. M. *Sur l'analyse numérique des methodes d'éléments finis hybrides et mixtes*. PhD thesis, Université Pierre-et-Marie Curie, Paris, 1977.

International Series of Numerical Mathematics, Vol. 114, © 1993 Birkhäuser Verlag Basel

Two-Dimensional Solute Transport

Alvaro L. Islas* David O. Lomen†

Abstract. This paper develops approximate solutions to a two-dimensional solute transport problem for either continuous or step inputs. The basic idea is to recognize that solute transport is mainly the result of two processes that can be treated independently, convection with the water flow and mixing by diffusion. Convection is treated exactly using the method of characteristics, and diffusion is accounted for with the method of singular perturbations. The integrals involved are no longer taken along vertical lines, as in one-dimensional problems, but now become contour integrals. This does not change the structure of the solution, it just makes the calculations more elaborate, and in some situations numerical algorithms are required to evaluate these integrals. Nevertheless, this method proves to be faster and less cumbersome than traditional difference and finite element schemes.

1 INTRODUCTION

There currently is a lot of interest in understanding the movement of pollutants and agricultural chemicals in soil water. Chemicals are added to the soil in the form of fertilizer, pesticides, etc., and removed by adsorption, dissolution, plant uptake and volatilization. Many people are concerned that undesirable chemicals might find their way into the groundwater. Growers are interested in making sure chemicals are present in the soil area adjacent to the plant roots in sufficient quantity to provide for optimum growth. However, the movement of soil water and the action of any dissolved chemical is a complicated system to model. Even with a precise mathematical model, the values of the physical and chemical constants it contains are rarely precisely known, and usually vary throughout the soil. Analytical studies of mathematical models are one way to develop an understanding of the process and determine the sensitivity of the movement of the chemicals to changes in the input parameters. The convection-diffusion partial differential equation is one such mathematical model. Reviews of analytical solutions of such equations are found in papers by Nielsen *et al.* [10] and van Genuchten and Jury [13]. Collections of analytical solutions have been given by van Genuchten and Alves [12], Javendel *et al.* [4], and Lomen [5]. Almost all of the analytical solutions mentioned above are for linear equations. Perturbation solutions of associated nonlinear partial differential equations in one spatial dimension are given in Lomen *et al.* [7] and Lomen *et al.* [6]. This paper considers a similar problem for two-dimensional flows, which are applied in situations with point or line sources of water and/or chemicals. The case for transport and diffusion of a pure solute (without plant uptake or reaction) will be covered in this paper in detail, and the process by which these effects can be included will be presented elsewhere (see Islas [2] for more details).

*Department of Mathematics, University of Arizona, Tucson, AZ 85721

†Department of Mathematics, University of Arizona, Tucson, AZ 85721, lomen@math.arizona.edu

2 GOVERNING EQUATIONS

Pure solute transport with a steady water flow is governed by the equation

$$\frac{\partial}{\partial t}(\theta c) = -\nabla \cdot (-D\theta \nabla c + \mathbf{q}c), \tag{2.1}$$

where $\nabla = (\partial/\partial x, \partial/\partial z)$ with x, z being the horizontal and vertical coordinates, respectively; t is time; θ is water content; c is the solute concentration; D is the hydrodynamic dispersion coefficient, and $\mathbf{q}$ is the flux rate.

We can simplify (2.1) by using the continuity equation,

$$\frac{\partial \theta}{\partial t} = -\nabla \cdot \mathbf{q}, \tag{2.2}$$

to obtain

$$\theta \frac{\partial c}{\partial t} = \nabla \cdot (D\theta \nabla c) - \mathbf{q} \cdot \nabla c. \tag{2.3}$$

2.1 Dimensionless Variables

It is advantageous to make the following change of variables:

$$x' = \frac{\alpha}{2}x, \qquad z' = \frac{\alpha}{2}z,$$
$$t' = \frac{\alpha k}{4}t, \qquad D' = \frac{\alpha}{k}D.$$

For the particular water flow of interest to us, that due to a line source along the y-axis (Lomen and Warrick [9]), θ and $\mathbf{q}$ are given by

$$\theta(x,z) = \frac{\alpha}{k}\phi(x,z), \qquad \mathbf{q}(x,z) = \nabla\phi - \alpha\phi\hat{e}_z,$$

with

$$\begin{aligned} \phi(x,z) &= 2e^{2z}\left(f(x,z) - \int_z^\infty f(x,p)\,dp\right), \\ f(x,z) &= e^{-z}K_0(\sqrt{x^2+z^2}), \end{aligned}$$

where K_0 is the Bessel function of the second kind of order zero and $\hat{e}_z$ is the unit vector in the z-direction.

In terms of the new spatial variables (which we now write without the primes), (2.3) becomes

$$\phi \frac{\partial c}{\partial t} = \nabla \cdot (D\phi \nabla c) - (\nabla\phi - 2\phi\hat{e}_z) \cdot \nabla c. \tag{2.4}$$

3 THE MATHEMATICAL APPROACH

To simplify the mathematical analysis, we view convection and diffusion as two independent processes. In (2.4), the first term in the right hand side represents mixing by diffusion, while the second gives convection with the water flow. Convection can be treated exactly with an application of the method of characteristics, and diffusion effects can then be included using the method of singular perturbations.

3.1 Convection

To treat convection, set $D = 0$ and rewrite (2.4) as

$$\frac{\partial c}{\partial t} + \mathbf{Q} \cdot \nabla c = 0, \tag{3.1}$$

where

$$\mathbf{Q} = \frac{\nabla \phi}{\phi} - 2\hat{e}_z.$$

We then apply the method of characteristics to solve this equation.

To leading order, the two-dimensional behavior is the same as in one dimension with the difference that the solute does not move anymore along vertical lines but along the streamlines (characteristic curves or just characteristics) given by $\mathbf{Q}(x, z) = (Q_1(x, z), Q_2(x, z))$ (vertical lines were the characteristics in the one-dimensional case). To see this, introduce the variable s that measures the distance along the streamlines (arclength); then the derivative with respect to s is the directional derivative in the direction of $\hat{\mathbf{Q}}$, the unit vector in the direction of $\mathbf{Q}$; i.e.,

$$\frac{\partial}{\partial s} = \hat{\mathbf{Q}} \cdot \nabla.$$

Since all quantities are now functions of s, then

$$\frac{\partial}{\partial s} = \frac{dx}{ds}\frac{\partial}{\partial x} + \frac{dz}{ds}\frac{\partial}{\partial z}.$$

Combining these equations produces the following set of equations for x and z in terms of s (also known as the characteristic equations):

$$\frac{dx}{ds} = \frac{Q_1(x, z)}{v(x, z)}, \qquad \frac{dz}{ds} = \frac{Q_2(x, z)}{v(x, z)}, \tag{3.2}$$

where $v = |\mathbf{Q}|$. Now, we can rewrite (3.1) in terms of s as

$$\frac{\partial c_0}{\partial t} + v\frac{\partial c_0}{\partial s} = 0, \tag{3.3}$$

where $v = v(s) = |\mathbf{Q}(x(s), z(s))|$ and $c_0 = c_0(s, t)$. The solution to (3.3) is found to be (Lomen *et al.* [8])

$$c_0(s, t) = c_0(0, t - f_1(s)), \qquad f_1(s) = \int_0^s \frac{ds'}{v(s')}.$$

Since we are interested in finding the solution in terms of x and z, it is appropriate to write f_1 as a function of (x, z). From the definition of arclength, $ds^2 = dx^2 + dz^2$, and from (3.2) it follows that

$$x'(z) = \frac{dx}{dz} = \frac{Q_1}{Q_2}, \qquad x(z_0) = x_0. \tag{3.4}$$

Now, f_1 can be rewritten as follows:

$$\begin{aligned} f_1(x_0, z_0) &= \int_0^{z_0} \frac{\sqrt{1 + x'^2(z)}}{v(x(z), z)}\, dz \\ &= \int_0^{z_0} \frac{\sqrt{1 + (Q_1/Q_2)^2}}{\sqrt{Q_1^2 + Q_2^2}}\, dz \\ &= \int_0^{z_0} \frac{dz}{|Q_2(x(z), z)|}. \end{aligned}$$

We are left with the problem of solving (3.4) with appropriate initial conditions.

3.1.1 Numerical Approximation of f_1. The special form of Q_1 and Q_2 precludes an analytic solution of (3.4). Thus, we have to solve (3.4) numerically, and the idea is as follows. Suppose we want the solution c_0 at (x_0, z_0, t_0). First, solve (3.4) backwards until $x = 0$. Note that, except for the origin from which streamlines radiate, one and only one streamline passes through each point. Save the solution at N points, N odd,

$$x_n = x(z_n), \qquad z_1 = 0 < z_2 < \cdots < z_N = z_0,$$

and then evaluate $f_1(x(z_0), z_0)$ using Simpson's Rule:

$$f_1(x_0, z_0) \approx \frac{\delta z}{3}(F_1 + 4F_2 + 2F_3 + 4F_4 + \cdots + 2F_{N-2} + 4F_{N-1} + F_N),$$

where

$$F_n = \frac{1}{|Q_2(x(z_n), z_n)|}.$$

Finally, the solution is given by

$$c_0(x_0, z_0, t_0) = c_0(0, 0, t_0 - f_1(x_0, z_0)).$$

Figure 1 shows the fixed streamlines and concentration fronts as they evolve in time.

3.2 Diffusion

The effects due to diffusion can be incorporated using the method of singular perturbations. First we assume that mixing occurs mainly in the direction of the flow (i.e, along the characteristics) and only around the front. Everywhere else, mixing is negligible. With this in mind, then the idea is to introduce a new variable that would amplify the region around the front, neglect higher order terms, and solve the resulting equation. Here we will sketch these steps; for more details see Islas [2].

3.2.1 Change of Variables. In order to apply this idea, we define a new coordinate system that would naturally adapt to this problem. Clearly it is constructed out of the streamlines (parameterized by the arclength s) and a system of curves perpendicular to them, defined by ψ (see Figure 2).

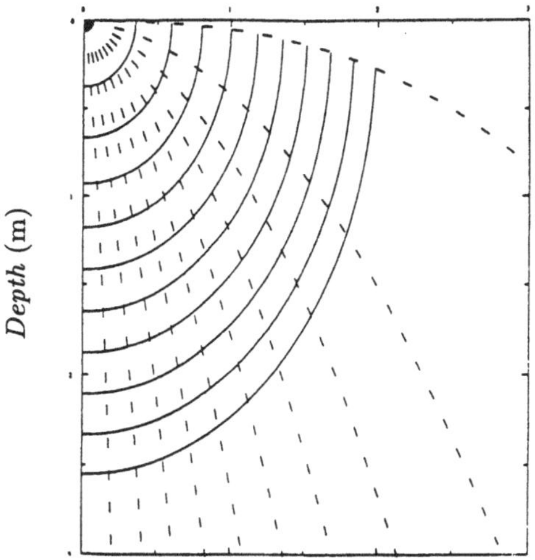

Figure 1: Concentration Fronts (solid lines) with Streamlines (dashed lines).

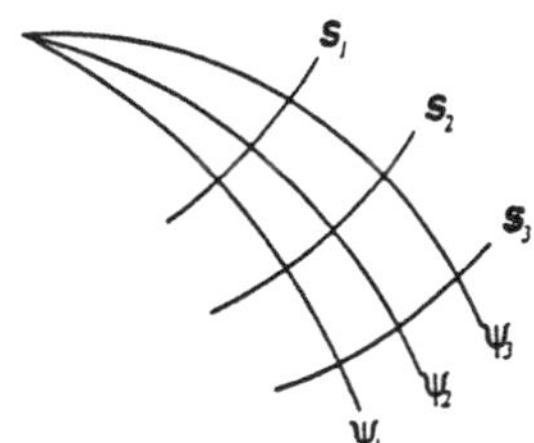

Figure 2: The New Coordinate System

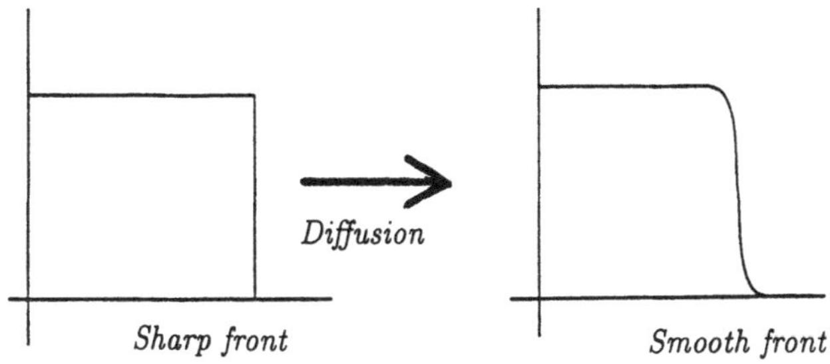

Figure 3: The Smoothing Effect of Diffusion.

Since this is an orthogonal system and is fixed in space (the streamlines do not change in time), (2.3) is essentially unchanged. We write (2.3) as

$$\theta\left(\frac{\partial c}{\partial t}+v\frac{\partial c}{\partial s}\right)=\nabla_{s,\psi}\cdot\left(\theta D\nabla_{s,\psi}c\right),$$

where the gradient and divergence operators are defined as follows:

$$\nabla_{s,\psi}c=\frac{\partial c}{\partial s}\hat{e}_s+\frac{\partial c}{\partial\psi}\hat{e}_\psi,\qquad \nabla_{s,\psi}\cdot\mathbf{A}=\frac{\partial A_1}{\partial s}+\frac{\partial A_2}{\partial\psi}.$$

Next, define the amplifying variable ξ and a new dependent variable U as follows:

$$\xi=\frac{s-s_f(t)}{D},\qquad U(\xi,\psi,t)=c(s_f(t)+D\xi,\psi,t),$$

where $s_f(t)$ denotes the location of the front along this particular streamline. After substitution, if we expand in powers of D, and keep only the lowest order terms ($U=u+$ higher order terms), we obtain the following equation:

$$\frac{\partial}{\partial\xi}\left(\theta\frac{\partial u}{\partial\xi}\right)-\left(v-\frac{ds_f}{dt}\right)\theta\frac{\partial u}{\partial\xi}=0. \tag{3.5}$$

In order to solve (3.5), we first need to determine the location of the discontinuities, and for that we need an equation for $s_f(t)$. This can be found from the definition of travel time. Let $s_f(t_0)$ be the location of the front at time t_0. Then, from Raats [11],

$$t-t_0=\int_{s_f(t_0)}^{s_f(t)}\frac{ds_f}{v(s)},$$

which gives an integral equation for $s_f(t)$. To obtain a differential equation, we just take the derivative with respect to t and find that

$$\frac{ds_f}{dt}=v(s_f(t)). \tag{3.6}$$

Now, we can proceed to solve (3.5) for u. Integrating (3.5) once gives

$$\theta\frac{\partial u}{\partial\xi}=Ae^{\int\left(v-s_f'(t)\right)d\xi}, \tag{3.7}$$

with

$$s_f'(t)=\frac{ds_f}{dt}$$

and A a constant of integration. Since at this point we are interested in an approximate solution, we use (3.6) to see that

$$\begin{aligned} v(s_f(t)+D\xi) &\approx v(s_f(t))+D\xi v'(s_f(t)) \\ &= s_f'(t)+D\xi s_f''(t)/s_f'(t), \end{aligned} \tag{3.8}$$

$$\theta(s_f(t)+D\xi)\approx\theta(s_f(t)), \tag{3.9}$$

with

$$v'(s_f(t)) = \left.\frac{dv}{ds}\right|_{s=s_f(t)}, \qquad s_f'' = \frac{d^2 s_f}{dt^2}.$$

Substituting (3.9) into (3.7) and dividing by θ leads to

$$\frac{\partial u}{\partial \xi} = \frac{A}{\theta} e^{\frac{1}{2} D s_f''(t)/s_f'(t)\xi^2},$$

which, when integrated once more, gives the inner solution u as

$$u(\xi, t) = \frac{A}{\theta} \int e^{\frac{1}{2} D s_f''(t)/s_f'(t)\xi^2}\, d\xi + B,$$

where B is another constant of integration.

To determine the constants of integration, we match the values of the inner solution u and the outer solution c at the front, $s = s_f(t)$, where we have the matching conditions

$$u(-\infty, t) = c(s_f(t), t), \qquad u(+\infty, t) = 0.$$

Then, the inner solution on the front is given in terms of s by

$$c(s, \psi, t) = \frac{1}{2} c_0(s_f(t), t) \operatorname{erfc}\left[\left(\frac{1}{2} D s_f''(t)/s_f'(t)\right)^{1/2} (s - s_f(t))\right],$$

where $c_0(s_f(t), t)$ is the convective solution evaluated at the front and $\operatorname{erfc}(x)$ is the complementary error function defined by

$$\operatorname{erfc}(x) = \frac{2}{\sqrt{\pi}} \int_x^\infty e^{-w^2}\, dw.$$

So, we have the convective solution being multiplied by a smoothing factor which affects it only around the front. This can be seen in Figure 3, where the sharp front has been smoothed at the corners.

4 CONCLUSIONS

We have sketched the methodology to find solutions for the transport equation. It is based on the assumption that convection and diffusion processes can be seen as occurring independently and therefore can be treated separately with the aid of different methods, the method of characteristics for the convection part and the method of singular perturbation for the diffusion one. Of crucial importance in applying this method is the availability of analytical solutions of the equation governing the underlying water flow. Thus, the search for analytical solutions of the equations governing water flow continues and new advances have been made in the one-dimensional case (see Broadbridge and White [1]) which has led to new analytical solutions of the transport equation (see Islas and Illangasekare [3]). These new solutions, although based on the same ideas, have the advantage of being given by closed-form formulae, not seen in previous works. Interesting situations occur when sinks and reactive chemicals are considered (see Islas [2]). The basic idea is the same so long as the streamlines remained fixed and a new coordinate system can be defined.

REFERENCES

[1] Broadbridge P., White I. Constant rate rainfall infiltration: a versatile nonlinear model. 1. Analytical solution. *Water Resour. Res.*, 24(1):145–154, 1988.

[2] Islas A. L. *Solute transport in several dimensions under steady water flows with root uptake and nonlinear chemical reactions.* PhD thesis, University of Arizona, 1993.

[3] Islas A. L., Illangasekare T. Solute transport under a constant rate infiltration. Analytical solution. *Water Resour. Res.* Submitted, 1993.

[4] Javendel I., Doughty C., Tsang C. F. *Groundwater Transport: Handbook of Mathematical Models.* Water Resources Monograph. Amer. Geophys. Union, Washington, 1984.

[5] Lomen D. O. Analytical solutions to partial differential equations modeling solute movement in porous media. Technical Report App. 110, University of Arizona, 1991.

[6] Lomen D. O., Islas A. L., Fan X., Warrick A. W. A perturbation solution for nonlinear solute transport in porous media. *Transport in Porous Media*, 6:739–744, 1991.

[7] Lomen D. O., Islas A. L., Warrick A. W. A perturbation solution for transport and diffusion of a single reactive chemical with nonlinear rate loss. In *Field-Scale Solute and Water Transport Through Soil*, pages 281–288. Birkhäuser Verlag, Basel, 1990. K. Roth, H. Flühler, W. A. Jury, and J. C. Parker, eds.

[8] Lomen D. O., Tonellato P. J., Warrick A. W. Salt and water transport in unsaturated soil for non-conservative systems. *Agr. Wat. Man.*, 8:397–409, 1984.

[9] Lomen D. O., Warrick A. W. Time-dependent linearized moisture flow solutions for surface sources. In *System Simulation in Water Resources*, pages 169–177. North-Holland Publishing Company, 1976. G. C. Vansteenkiste, ed.

[10] Nielsen D. R., van Genuchten M. Th., Biggar J. W. Water flow and solute transport processes in the unsaturated zone. *Water Resour. Res.*, 22:895–1085, 1986.

[11] Raats P. A. C. Convective transport of solutes by steady flows. I. General theory. *Agr. Wat. Man.*, 1:201–218, 1978.

[12] van Genuchten M. Th., Alves W. J. Analytical solutions of the one-dimensional convective-dispersive solute transport equation. Technical Bulletin 1661, U.S. Department of Agriculture, Agricultural Research Service, Washington, D.C., 1982.

[13] van Genuchten M. Th., Jury W. A. Progress in unsaturated flow and transport modeling. *Rev. of Geophys.*, 25:135–140, 1987.

International Series of Numerical Mathematics, Vol. 114, © 1993 Birkhäuser Verlag Basel

Multiphase Saturation Equations, Change of Type and Inaccessible Regions

Barbara Lee Keyfitz*

Abstract. We identify a class of flux functions which give rise to conservation laws which are hyperbolic except along a codimension one subspace of state space. We show that a number of systems modelling porous medium flow can be regarded as perturbations of such systems, and describe the phenomenon of change of type for these perturbations. We also discuss a property of solutions of such systems, the existence of inaccessible regions - subsets of state space which appear to be avoided by solutions.

1 INTRODUCTION

Equations that change type appear in some models for multiphase flow, where they cause a certain amount of controversy. The context is as follows. In a fluid consisting of several phases, components or both, in a flow regime where dissipative, diffusive or dispersive effects should be negligible, one ends up with a system of quasilinear first-order partial differential equations, in space and time, which are expected to be of hyperbolic type. Details of the modelling depend on the specific problem — Darcy's law for porous medium flow, the usual conservation of mass and momentum for compressible two-phase flow, other models for transport of solutes or sediments — but the same disconcerting behavior occurs: not only do the characteristic speeds depend on the state variables (as is usual) but also there is a region of state space where some characteristic speeds are not real-valued. In the simplest case, where there are two equations only and a single space variable, this means that the system changes type from hyperbolic to elliptic. Recall that steady transonic flow also contains both supersonic (hyperbolic) and subsonic (elliptic) regimes. However, it turns out that the two sorts of problems — steady and unsteady models — have different mathematical structure; see Keyfitz [8]. In addition, issues that are important for applications — such as well-posedness of the initial-value problem — are different in the two cases. In this paper, 'change of type' refers to systems of first-order quasilinear equations, modelling unsteady problems, in which some characteristic speeds change from real to complex in a region of state space. We shall assume that the equations are in conservation form.

More background on change of type in porous medium flow can be found in the review article by Keyfitz [10].

In the next section, we describe some models where the phenomenon appears. Based on these examples, we identify a class of nonlinear flux functions with the property that the associated conservation laws are hyperbolic everywhere, but nonstrictly hyperbolic on a codimension one submanifold of phase space. For systems of two equations, we write down some explicit criteria for membership in this class.

*Mathematics Department, University of Houston, Houston, Texas 77204-3476, keyfitz@uh.edu

In §3 what happens when an equation in this class is subjected to a general perturbation is discussed: change of type may occur. In the case of two equations, for example, elliptic regions arise near the original curve of nonstrict hyperbolicity. Other phenomena affecting the structure of solutions of conservation laws, such as curves of linear degeneracy, are also associated with the perturbations.

The next section, §4, contains an informal discussion of one aspect of the ill-posedness associated with change of type: the existence of so-called inaccessible regions in the flow. Difficulties associated with this are mentioned in Allen *et al.* [1]. We show that this behavior is similar to strictly hyperbolic systems which admit linear degeneracies.

2 MULTICOMPONENT SATURATION EQUATIONS

The general form of the equations we study is

$$\partial_t p_i + \partial_x (p_i v_i(P)) = 0, \quad 1 \le i \le n. \tag{2.1}$$

Here $P = (p_1, \ldots, p_n)$ is the vector of states, and we define $F = (f_1, \ldots, f_n)$, the vector of corresponding fluxes. The particular form $f_i = p_i v_i(P)$ is appropriate when each p_i is the density or relative saturation of a component or phase. Then (2.1) is the set of continuity equations for n species, and v_i, the velocity of the ith species, depends on the state vector.

A system of kinematic equations like (2.1) is part of a more complete model of a fluid system. Momentum and energy equations have been omitted and the system has been closed by the assumption that velocities depend on densities or concentrations alone. Rarely is this completely realistic. However, it is an approximation which is often taken seriously. For example, it is used to describe the so-called *miscible displacement problem* in enhanced oil recovery, in which one component of P represents the saturation of a solvent in a fluid whose concentration is given by the other component. In this case it is often assumed that all the v_i are identical (see, for example, Johansen and Winther [5]). Another example arises in three-phase *immiscible* porous medium flow, where the momentum equation is replaced by Darcy's law and it is further assumed, for a single space dimension, that the pressure equation can be solved explicitly. In this case, the v_i are complicated functions of the phase fractions, P, involving the three-phase relative permeabilities, which are usually determined by interpolation (see Allen *et al.* [1]). One classic example of a kinematic equation is a continuum model for traffic flow; the adaptation of this model to a system purportedly describing two-directional traffic flow leads to change of type, (Bick and Newell [2]).

Standard models for two-phase compressible, nonreacting flow consist of a pair of equations which represent conservation of each phase, coupled with momentum and energy transfer equations; see Stewart and Wendroff [18]. These equations reduce asymptotically to a pair of continuity equations of the form (2.1) when the faster-moving waves are ignored; see Keyfitz [9].

Finally, kinematic equations like (2.1) are important in some chemically reacting systems, such as chromatography and other adsorption processes; see Rhee *et al.* [15] and Temple [20]. Change of type and failure of strict hyperbolicity occur in some but not all of these systems.

We examine systems with the property that a single pair of eigenvalues coincides for some values of P, while remaining real everywhere. For concreteness, consider a pair of equations,

$$\begin{aligned} p_t + f_x &= 0, \\ q_t + g_x &= 0, \end{aligned} \tag{2.2}$$

with Jacobian matrix

$$dF = \begin{pmatrix} f_p & f_q \\ g_p & g_q \end{pmatrix}.$$

The eigenvalues of dF, which are the characteristic speeds, are real or complex according to the sign of the discriminant

$$D(p,q) = \left(\frac{f_p - g_q}{2}\right)^2 + f_q g_p. \tag{2.3}$$

System (2.2) is *strictly* hyperbolic when D is positive, and, since this is an open condition, (2.2) has this property on an open subset of $\mathbf{R}^2$. For the same reason, if (2.2) is strictly hyperbolic for a flux vector F_0 in the entire region of physical interest (the positive quadrant or unit square, say), then it remains strictly hyperbolic in this entire region for all perturbations of F_0 in a sufficiently small C^1-open set about F_0.

However, it may happen that D is simply nonnegative everywhere. In this case, the system (2.1) or (2.2) is called *nonstrictly* hyperbolic. The condition $D \geq 0$ is *not* open: if there are no other constraints on F there will be points (p,q) with $D < 0$ in any neighborhood of a point where $D = 0$. Let us introduce a terminology for flux functions which satisfy a constraint which prevents this.

Definition 2.1 *We say that F is a* nonstrictly hyperbolic (NSH) flux function *for system (2.2) in a subset $R \subset \mathbf{R}^2$ if the discriminant D defined in (2.3) is nonnegative for all $(p,q) \in R$ and $D = 0$ for at least one point in the interior of R. The* coincidence locus, *Σ, is the set of points (p,q) where $D = 0$.*

The definition excludes strictly hyperbolic systems and also systems, such as the gas dynamics equations, which lose strict hyperbolicity at the vacuum state, on the boundary of the physically interesting region.

One well-known class of NSH flux functions consists of gradients of a potential; if $F = \nabla\Phi$ then $dF = d^2\Phi$ is symmetric and hence its eigenvalues are real. In this case,

$$D = \left(\frac{\Phi_{pp} - \Phi_{qq}}{2}\right)^2 + \Phi_{pq}^2,$$

and the system is NSH if there is a point (p,q) where the two equations $\Phi_{pp} - \Phi_{qq} = 0$ and $\Phi_{pq} = 0$ are satisfied. If Φ is an arbitrary C^2 function of two variables, then these equations constitute two independent conditions, and Σ consists of isolated points (p,q) — usually called *umbilic* points. Nonstrictly hyperbolic fluxes with umbilic points have been studied extensively, beginning with Schaeffer and Shearer [16].

However, many examples that arise in modelling have a different structure: they are nonstrictly hyperbolic, but the condition $D = 0$ reduces to a single equation whose solution space Σ is a curve contained in R. An example is the class of separated potentials: a flux $F = (f(p), g(q))$ corresponding to an uncoupled system has this structure, as does the flux that results in a triangular Jacobian:

$$F = (f(p,q), g(q)). \tag{2.4}$$

In this example, eigenvalue coincidence occurs where

$$G(p,q) \equiv f_p(p,q) - g'(q) = 0,$$

and if this equation is satisfied at a point (p_0, q_0) then, by the implicit function theorem, it is satisfied on a curve through that point provided

$$\nabla G(p_0, q_0) \neq 0. \tag{2.5}$$

A two-component flow with a flux of the form (2.4) has the property that the velocity of the q-component is independent of the density of the p-component: the system is at least partially uncoupled. This may not be a realistic approximation for multiphase flows; though the two-way traffic equations, (Bick and Newell [2]), have this property when there is no interaction between the two directions of flow.

A more interesting example of a NSH flux is given by the two-component miscible displacement saturation equations. These can be written with a flux vector of the form

$$F(p,q) = (p\phi(p,q), q\phi(p,q)) \tag{2.6}$$

(see Johansen and Winther [5]); in this case

$$D = \left(\frac{p\phi_p + q\phi_q}{2}\right)^2,$$

and the system is nonstrictly hyperbolic if the equation $G \equiv p\phi_p + q\phi_q = 0$ has a solution (p_0, q_0); eigenvalue coincidence occurs along a curve under the nondegeneracy condition (2.5). A two-fluid model for gas chromatography, (Temple [20]), has a structure similar to (2.6), except that the second component of F is multiplied by a constant; however, in the standard model (using the Langmuir isotherm), there is no eigenvalue coincidence: $D > 0$ everywhere in the physical region.

The two NSH models above — a flow in which the velocities of both components are the same, and a flow in which one is independent of the other — seem to represent extremes in kinematic modelling. However, they have similar mathematical properties. Furthermore, general perturbations of either model, which take them out of the class of NSH systems, also have much in common, as we shall explore in the next section.

In the remainder of this section, we give a brief description of models with the property that Σ is a curve. This is motivated by the observation that special multiphase kinematic flows, as in the examples given above, appear to have this structure, rather than the umbilic structure of a NSH flux which derives from a potential. We shall refer to this class of NSH fluxes as *coincidence-line* fluxes.

In some multiphase saturation models, such as three-phase porous medium flow, the flux vectors must either be NSH or show change of type (Shearer and Trangenstein [17]), and this has motivated studying perturbations of umbilic points. Given a model with an elliptic region, one can embed it in a family of models in which the elliptic region is shrunk to a point or to a line, and it is to some extent a matter of taste which one chooses to do. For example, in Vinod [21], the elliptic region occurring in a quadratic model for the two-way traffic equations is shrunk to a line — the limit of zero interaction — while in Holden and Holden [3] an equivalent quadratic model is considered as a perturbation of an umbilic. Nonstrictly hyperbolic systems whose eigenvalues coincide along a line have simpler solutions than hyperbolic systems with an umbilic point. It would be interesting to be able to show that the constraints which force a NSH flux to be of coincidence line type have some physical basis.

We make the following obvious remark.

Proposition 2.1 *Let $F \in C^2$ be a NSH flux vector in the sense of Definition 2.1 and let D be the corresponding discriminant. Then at any point where $D = 0$ we also have $\nabla D = 0$.*

The converse holds as well under some additional conditions on F; for example, $F \in C^3$ and nondegeneracy conditions on d^2D. Informally, we might say that D needs to be a perfect square if the flux is to be NSH, and Proposition 2.1 shows that this condition is not likely to be met by an arbitrary pair of functions f and g. Furthermore, in order for a NSH flux vector to be a coincidence-line flux (rather than umbilic), the three equations $D = \nabla D = 0$ must be equivalent to a single equation in the two variables p and q, to yield a curve Σ.

The proposition does not suggest any useful way of characterizing these fluxes. But looking at the geometry of conservation laws in the plane gives some insight. The Jacobian matrix, dF, of a NSH flux vector has one real eigenvector at every point in R. Denote the eigenvector by $\zeta(U)$, where $U = (p, q)$, and suppose it can be chosen to depend smoothly on U. The integral curves of the line field generated by ζ give a foliation of R; if the curves are written in the form $\chi(U) = c$ then χ is a Riemann invariant.

Now, F is a mapping from R to a subset of $\mathbf{R}^2$, and by translating F by a constant vector (which does not affect the conservation law system) and possibly scaling by a constant factor, we can assume that $F(R) \subset R$. (For our purposes, we may assume R has compact closure, and then the additional assumption made here is that F is bounded.) In this case, the equation

$$(dF(U))\zeta(U) = \lambda(U)\zeta(U) \tag{2.7}$$

has the interpretation that F maps each Riemann invariant curve to a translate of that curve. (This follows from differentiating along the curve. At corresponding points U and $F(U)$, the tangent vectors, ζ, to the curve are parallel, as a consequence of equation (2.7).) Let us consider the special case that F respects the foliation: that is, the image of a Riemann invariant curve is a Riemann invariant curve. (This does not seem a particularly natural assumption, but it holds for a number of physical models, including the ones mentioned above.) Now, either F maps every curve to itself, or F maps each curve to a different curve. The condition on tangents implies that a Riemann invariant curve that is mapped to itself must be a straight line segment. On the other hand, if curves are mapped into each other by F, then the tangency condition means that all the curves are, effectively, translates of each other, so that χ is of the form $q - r(p)$, at least locally.

Thus we state

Proposition 2.2 *Let R be a bounded subset of $\mathbf{R}^2$, and let ζ be a smooth line field defined on R satisfying one of the two constraints: the integral curves of ζ are straight lines, or the integral curves are translates of a single curve. Then there is a smooth flux function, F, defined on R, with eigenvector $\zeta(U)$ at each point U.*

Proof: Temple [20] has shown how to construct a flux F corresponding to any line field with straight-line integral curves. Writing $\zeta = (1, h(U))$, the condition that ζ have straight-line integral curves is

$$\nabla h \cdot \zeta = 0,$$

so h is any smooth solution of $h_p + hh_q = 0$; given h, Temple shows that the form of F is (up to inessential normalizations)

$$F(U) = \phi(U)\begin{pmatrix} 1 \\ h(U) \end{pmatrix} + \begin{pmatrix} 0 \\ H(h) \end{pmatrix}. \tag{2.8}$$

Here ϕ is an arbitrary function of U and H an arbitrary function of h. Temple's construction begins with the fixed solution $h(U)$; however, it is clear that from any foliation of R by straight lines, one can construct ζ (up to the condition that the first component be nonzero), and F. This generalizes the flux of (2.6), for which ζ is just the field of radial lines: $\zeta = (1, q/p)$ with Temple's choice of normalization. The eigenvalue corresponding to ζ is $\lambda(U) = \phi_p + h\phi_q = \nabla\phi \cdot \zeta = \partial_\zeta \phi$. The other eigenvalue is $tr(dF) - \lambda = \phi h_q + H_q = (\phi + H'(h))h_q$.

When $\chi = q - r(p)$, then $\zeta = (1, r'(p))$; it is easy to verify that

$$F(U) = \begin{pmatrix} a(p) - q \\ b(p) \end{pmatrix} \tag{2.9}$$

is a flux vector which gives rise to this eigenvector as long as b, a and r are related by $b' = r'(a' + r')$. The corresponding eigenvalue is $\lambda = a' - r'$, and the other eigenvalue is r'. Notice that there is no condition imposed on r other than sufficient smoothness. The region R will, of course, depend on r. ∎

So far, we have not required that the fluxes be NSH. However, for both the families we have constructed, it is straightforward to impose the condition that eigenvalues coincide on a subset of R; under suitable nondegeneracy conditions, this occurs along a curve for both types of models.

Proposition 2.3 *Let a flux function F be defined by (2.8) or (2.9) on a domain R. Then F is a nonstrictly hyperbolic flux vector if there is a point (p_0, q_0) where, in the first case,*

$$G(p,q) \equiv \phi_p + h\phi_q - \phi h_q - H_q = 0,$$

and, in the second,

$$G(p) \equiv a'(p) + 2r'(p) = 0.$$

Furthermore, $U_0 = (p_0, q_0)$ lies on a nondegenerate curve, Σ, of coincident eigenvalues in the first case if $\nabla G(U_0) \neq 0$. In the second case, Σ includes the line $p = p_0$, and this is an isolated coincidence line if $G'(p_0) \neq 0$.

Proof: This is a straightforward calculation using the expressions for the eigenvalues given in Proposition 2.2. The existence and nondegeneracy of Σ follow from the implicit function theorem applied to G. ∎

Another interpretation of equation (2.7) is as follows. The flux corresponding to (2.4) produces a nondegenerate curve as a line of coincident eigenvalues as long as (2.5) holds. Again suppose that F maps the domain R into R, and take any smooth coordinate transformation, T, from R to itself; then the composition

$$\hat{F} = T \circ F \circ T^{-1} \tag{2.10}$$

defines a new flux function on R. (It is not particularly natural, on the basis of the physical problem, to regard F as a mapping of R to itself; however, it is a reasonable approach if one wants to study the structure of eigenvalues of dF.) Defining $V = T(U)$, one has another system

$$\partial_t V + \partial_x(\hat{F}(V)) = 0. \tag{2.11}$$

There need not be any relation between the eigenvalues of dF and those of $d\hat{F}$. In fact, defining $S = T^{-1}$ as an abbreviation,

$$d\hat{F}(V) = (dT)_{F\circ S(V)}(dF)_{S}(V)(dS)_V \tag{2.12}$$

where the subscripts on the right indicate where the Jacobians are to be evaluated. By contrast, if one begins with the original conservation law (2.1) and applies the change of coordinates to U, one obtains the quasilinear system

$$\partial_t V + [(dT)_V(dF)_U(dS)_V]\,\partial_x V = 0 \tag{2.13}$$

which cannot, in general, be put in conservation form. Comparing the Jacobian in (2.12) with the matrix in (2.13), we see that they are the same only if

$$(dT)_{F\circ S(V)} = (dT)_V. \tag{2.14}$$

In this case, (2.11) is the same as the original system, but written in new coordinates. (The two systems are not equivalent when it comes to weak solutions or shock structure, but they have the same characteristic speeds, and the corresponding eigenvectors transform to each other under T.) However, (2.14) is a very restrictive condition.

Nonetheless, one can generate conservation laws (2.11) with a structure that is qualitatively like a given one, as follows. Suppose that, with $U = (p,q)$, F is of the form (2.4), and so dF is upper triangular. In that case, so is $C(dF)$, where

$$C = \begin{pmatrix} 1 & e(U) \\ 0 & c(U) \end{pmatrix}$$

is an upper triangular matrix. If now we replace (2.14) by

$$(dT)_{F\circ S(V)} = (dT)_V C \tag{2.15}$$

then (2.11) is a conservation law whose eigenvalues are those of $C(dF)$ — that is f_p and cg_q. In particular, they are real everywhere in R and coincide along the curve Σ where $f_p(U) = c(U)g_q(U)$. Thus we have proved

Proposition 2.4 *Let F be any smooth flux of the form (2.4) which maps a subset R of $\mathbf{R}^2$ to itself. Let $T(U) = V$ be a nonsingular coordinate change in R and C any smooth upper triangular matrix. Then, provided the compatibility condition (2.15) is satisfied, the system (2.11), with flux defined by (2.10), is hyperbolic in R, with eigenvalues f_p and cg_q. Equation (2.11) has a NSH flux vector $\hat{F}$ if there is a point, U_0, where*

$$G(U) \equiv f_p(U) - c(U)g'(q) = 0,$$

and this equation is satisfied on a curve Σ if $\nabla G(U_0) \neq 0$.

This proposition relates several of the examples above — for example, the transformation from Cartesian to polar coordinates turns the triangular flux function, (2.4), into the flux vector for the miscible displacement problem, (2.6). In [11], we show that a flux generated this way produces a discriminant D which is a perfect square, as in Proposition 2.1. One class of such fluxes is given by

$$\hat{F}(u,v) = \varphi(u,v)\begin{pmatrix} t(Q) \\ r(Q) \end{pmatrix} + \begin{pmatrix} s(Q) \\ a(Q) \end{pmatrix}, \tag{2.16}$$

where φ is any function of $V = (u, v)$; t, r, s and a are arbitrary functions satisfying the constraint that

$$\begin{aligned} u &= pt(q) + s(q), \\ v &= pr(q) + a(q), \end{aligned}$$

is a locally invertible change of coordinates, and $Q(u, v)$ is the solution of

$$ur(Q) - vt(Q) = b(Q) \equiv r(Q)s(Q) - t(Q)a(Q). \tag{2.17}$$

Since any smooth solution of the quasilinear equation

$$t(b' - ur')Q_u + r(b' + vt')Q_v = 0$$

with $b(Q(0,0)) = 0$ satisfies (2.17) and so generates a flux of the form (2.16), this generalizes (2.8). Finally, we have expressions for the eigenvalues of the Jacobian $d\hat{F}$ (from Proposition 2.4):

$$\lambda_1 \equiv f_p = \varphi_u t + \varphi_v r$$

and

$$\lambda_2 \equiv cg_q = \frac{\varphi(r't - rt') + a't - s'r}{ur' - vt' - (rs + at)'},$$

which yield a nondegenerate coincidence line when Q is nonconstant. Derivation of (2.16) will be found in [11].

3 PERTURBATIONS OF NONSTRICTLY HYPERBOLIC FLUXES

Suppose F is a NSH flux with eigenvalue coincidence along Σ, a nondegenerate curve given by

$$\delta(p, q) = 0$$

where $\delta^2 = D$ and $\nabla\delta \neq 0$. (Here D is given by (2.3).) We consider a one-parameter family of perturbed fluxes

$$\tilde{F} = \begin{pmatrix} f + \epsilon\phi \\ g + \epsilon\psi \end{pmatrix}.$$

In general, the new flux will not be NSH; in fact (tildes denote the perturbed quantities), the discriminant of $\tilde{F}$ is

$$\tilde{D}(p,q) = D(p,q) + \epsilon P(p,q,\epsilon) = \delta^2 + \epsilon P$$

where

$$P(p,q,\epsilon) = \frac{(f_p - g_q)}{2}(\phi_p - \psi_q) + \phi_q g_p + \psi_p f_q + \epsilon\left[\left(\frac{\phi_p - \psi_q}{2}\right)^2 + \phi_q\psi_p\right].$$

Proposition 3.1 *Let $\tilde{F}$ be a smooth perturbation of a smooth coincidence-line NSH flux with $\Sigma = \{(p,q) \mid \delta(p,q) = 0\}$. Suppose that $P(p,q,0) > 0$ on Σ. Then, for sufficiently small $\epsilon > 0$, $\tilde{F}$ is strictly hyperbolic near Σ, while for $\epsilon < 0$ there will be a nonhyperbolic strip near Σ whose width is of order $\sqrt{\epsilon}$. On the other hand, if P changes sign on Σ, let*

$$P(p(\tau), q(\tau), 0) = \tau\hat{P}(p(\tau), q(\tau)) \tag{3.1}$$

where τ parameterizes Σ and $\hat{P} > 0$ there. Then, $d\tilde{F}$ has nonreal eigenvalues inside a region whose boundary is, to a first approximation, a narrow parabola with its vertex near $U_0 = (p(0), q(0))$, opening toward $\tau > 0$ or $\tau < 0$ when $\epsilon < 0$ or $\epsilon > 0$, respectively.

Proof: The result when $P \neq 0$ on Σ follows immediately from applying the implicit function theorem to the two equations $\delta \pm \sqrt{-\epsilon P} = 0$: if $\epsilon < 0$, there are two solution curves for small ϵ and $\tilde{D}$ is negative between them.

If P changes sign along Σ, then equation (3.1) implies that the change of sign is nondegenerate. One can now solve

$$\delta(p(\tau,\epsilon), q(\tau,\epsilon)) \pm \sqrt{-\epsilon\tau\hat{P}} = 0$$

for each fixed τ, again using the implicit function theorem, since $\nabla\delta \neq 0$. The geometry of the nonhyperbolic region follows. ■

Perturbations of this type also have a connection with genuine nonlinearity. Curves of linear degeneracy (corresponding to isolated local extrema of the characteristic speeds along the eigenvectors of dF) appear when certain nonstrictly hyperbolic systems are perturbed so that they become strictly hyperbolic [6]. For these systems, the solution of the Riemann problem for the nonstrictly hyperbolic equation is qualitatively like the solution of the perturbed, strictly hyperbolic problem, which approaches it in the limit as $\epsilon \to 0$.

Curves of linear degeneracy also bifurcate from distinguished points on the boundary between an elliptic and a hyperbolic region in state space [7]. This bifurcation may occur generically at points like U_0 in Proposition 3.1. As we shall discuss in the next sectio, there is also some resemblence between the solutions in the two cases corresponding to $\epsilon > 0$ and $\epsilon < 0$ of Proposition 3.1.

4 INACCESSIBLE REGIONS

A disturbing feature of change of type in saturation equations like (2.1) is that there appear to be open sets in phase space, corresponding to physically feasible saturation vectors, which are nevertheless inaccessible because the Jacobian is nonhyperbolic there, (Allen *et al.* [1]). There are two related questions here. The first is whether some indefensible assumption in the physical model has resulted in a saturation vector which is linearly unstable. Unlike simplified models for phase transitions which change type and are nonhyperbolic precisely for the physically unstable range of the order parameter (see Pego and Serre [14] for an example), the flows discussed in this paper are not expected to contain unstable states.

The second point is a mathematical one. The nonhyperbolic region has some special properties. Specifically, solutions to the Riemann problem, when the data are in the hyperbolic region, avoid the nonhyperbolic states. This has been demonstrated analytically for some systems and is believed on the basis of numerical experience for others (see Allen *et al.* [1] and Pego and Serre [14]); there are no counterexamples, to the best of my knowledge. On the other hand, for more general initial data (Cauchy data) the hyperbolic region is not invariant: data in the hyperbolic region for which the solution enters the nonhyperbolic region are given in Holden *et al.* [4] and Pego and Serre [14]. The fact that Riemann data and Cauchy data behave so differently raises additional questions, which we leave aside.

In this section, we try to shed a bit of light on the second, mathematical, question by calling attention to an analogous phenomenon in strictly hyperbolic systems which contain hypersurfaces (curves, in the case of two equations) of linear degeneracy in state space. We shall concentrate on Riemann problems, since this is where the behavior is seen in flows which change type. There are some implications about the modelling of flows, because the analogy, even if only in Riemann problems, with a strictly hyperbolic system suggests strongly that the appearance of inaccessible regions is linked more to wrinkles in the nonlinear dependence of the fluxes than it is to change of type in the equations. The fact that both examples occur as perturbations of coincidence-line fluxes suggests that there may be a relationship between them.

Riemann solutions for conservation laws without convexity (genuine nonlinearity) assumptions were first given by Liu [12] by constructing a solution separately in each wave family and superimposing waves from different families. We formulate a result for a scalar equation, where we can give an explicit description, and then indicate the generalization.

Definition 4.1 *Let $u_t + f(u)_x = 0$ be a scalar conservation law. We say the Riemann problem with data $\{a, b\}$* avoids states *in a nonempty subset $\Sigma \subset (a, b)$ if neither the solution to the Riemann problem*

$$u(x,0) = \begin{cases} a, & x < 0, \\ b, & x \geq 0, \end{cases} \tag{4.1}$$

nor the solution to

$$u(x,0) = \begin{cases} b, & x < 0, \\ a, & x \geq 0, \end{cases} \tag{4.2}$$

takes values in Σ.

We have the following result.

Proposition 4.1 *Let $f \in C^1$, and let f' be strictly monotone on open intervals with no accumulation point. If f is strictly convex (or concave) on $[a,b]$, then no states in (a,b) are avoided. However, any interval on which f' is not weakly monotone will contain at least one subinterval of avoided states.*

Proof: Osher [13] presents a formula which gives a closed-form solution to the Riemann problem. It is equivalent to the following construction.

The *lower convex hull* of f on $[a,b]$ is

$$\underline{f}(u;a,b) = \inf\left\{f(c) + (u-c)\frac{f(d)-f(c)}{d-c}\right\}$$

where the *inf* is taken over c and d with $a \leq c \leq u \leq d \leq b$. Similarly the *upper convex hull* of f on $[a,b]$ is

$$\overline{f}(u;a,b) = \sup\left\{f(c) + (u-c)\frac{f(d)-f(c)}{d-c}\right\}$$

with the *sup* taken over the same domain. Write the Riemann problem as

$$u(x,0) = \begin{cases} u_L, & x < 0, \\ u_R, & x \geq 0. \end{cases}$$

For problem (4.1), where $u_L < u_R$, solve the Riemann problem for $u_t + \underline{f}(u)_x = 0$. For problem (4.2), with $u_L > u_R$, use $u_t + \overline{f}(u)_x = 0$. The Riemann problem for a (nonstrictly) convex function f is solved as follows. The centered solution $u(\zeta)$, $\zeta = x/t$, satisfies

$$[-\zeta + f'(u(\zeta))]\, u'(\zeta) = 0$$

in the sense of distributions, and can be written

$$u(\zeta) = g^{-1}(\zeta), \tag{4.3}$$

where $g \equiv f'$ is monotone. If g is constant on an interval I, then g^{-1} is discontinuous and the interior of I is not in its range. If g is increasing then (4.3) provides a solution for $u_L < u_R$; if g is decreasing, then (4.3) is a solution with $u_L > u_R$.

If f is strictly convex on (a,b), then $\underline{f} = f$ on (a,b) and g is strictly increasing. Then (4.3) yields a continuous solution (rarefaction wave) if $u_L < u_R$, and the range of g is the entire interval $[u_L, u_R]$. (In this case, $\overline{f}$ is a straight line segment

$$\overline{f}(u) = f(a) + (u-a)\frac{f(b)-f(a)}{b-a} \equiv f(a) + (u-a)s,$$

and $\overline{f}' = \overline{g} \equiv s$ in (a,b). The solution to the Riemann problem with $u_L = b$ and $u_R = a$ is $u(\zeta) = \overline{g}^{-1}(\zeta)$: u is piecewise constant with a discontinuity at $\zeta = s$. The range of $\overline{g}^{-1}$ is the two values $\{u_L, u_R\}$.)

If f is strictly concave, then $\overline{f} = f$ and $\underline{f}$ is a line segment, and the result is the same: this time $f' = g$ is strictly decreasing and has a continuous inverse, and no states are avoided.

Finally, if f is neither convex nor concave, then both $\overline{f}$ and $\underline{f}$ differ from f, and both contain line segments. In fact, let there be a point c in (a,b) such that f' is strictly increasing on $(c-\epsilon,c)$ and strictly decreasing on $(c,c+\epsilon)$ for some $\epsilon>0$. Then f is strictly convex on the first interval and strictly concave on the second, so if $a<c-\epsilon$ and $b>c+\epsilon$ then there is an open interval $(c-\delta,c+\delta)$, $0<\delta<\epsilon$, on which $\underline{f}<f<\overline{f}$. From the construction of the convex hulls, $\underline{f}$ and $\overline{f}$ are affine functions on $(c-\delta,c+\delta)$, and hence this interval is not in the range either of $\overline{g}^{-1}$ or of $\underline{g}^{-1}$. ∎

For a system of conservation laws, Liu [12] constructs a curve $\gamma(U_L)$ in state space which is locally a shock (part of the Hugoniot locus) or a rarefaction. This provides a higher-dimensional analogue to $\underline{f}$ and $\overline{f}$ and a solution like the scalar one can be constructed. The construction works whenever F has isolated hypersurfaces of linear degeneracy (which play the same role as the isolated extrema of f') and the eigenvalues of dF are separated. If $U_M\in\gamma_1(U_L)$ for a 1-wave curve, say, then some subintervals of that curve will not be in the range of the Riemann solution. Further, for U_R in a neighborhood of such a U_M, there will continue to be open sets of the interval that are missed in solving the Riemann problem for $\{U_L,U_R\}$. It is no longer the case that $U_M\in\gamma_1(U_L)\Rightarrow U_L\in\gamma_1(U_M)$. However, for small-amplitude waves, U_L will be near $\gamma(U_M)$, and a generalization of the idea of *avoided states* can be given in several ways. One formulation is to fix a left state U_L and consider right states in an open ball $\mathcal{B}$ of radius r centered at a point U_0. For each $U_R\in\mathcal{B}$, let $\mathcal{S}(U_R)\subset\mathbf{R}^n$ be the range of the Riemann solution with data $\{U_L,U_R\}$, and define

$$\mathcal{S}(\mathcal{B})=\bigcup_{U_R\in\mathcal{B}}\mathcal{S}(U_R).$$

If F is genuinely nonlinear, $\mathcal{S}(\mathcal{B})$ has at most $n+1$ connected components. Also, if U_0 is sufficiently close to U_L and F is genuinely nonlinear at U_L, then $\mathcal{S}(\mathcal{B})$ has at most $n+1$ components. For a system with linear degeneracies, as $|U_0-U_L|$ grows, $\mathcal{S}(\mathcal{B})$ will develop more than $n+1$ components at some U_0. The introduction of new components into $\mathcal{S}(\mathcal{B})$ corresponds to production of additional discontinuities in g^{-1} in the scalar case. Thus it is associated with avoided states. The qualitative change in $\mathcal{S}(\mathcal{B})$ also suggests a lack of continuous dependence of the solution on the data.

Avoided states differ from the inaccessible regions in systems that change type, as they are not precisely the complements of invariant regions. Similar behavior is observed in the two cases, since sampling data repeatedly in the exterior of a convex set, E, produces a solution which never enters a convex set C contained in E. The difference is that for Riemann data in systems which change type, one can apparently take E to be the elliptic region and C to coincide with E. For hyperbolic equations with linear degeneracies, C is strictly smaller than E. Nonetheless, the similarities are striking.

In a recent paper, Temple [19] contrasts Riemann problems (their relation to stability and asymptotics) in genuinely nonlinear strictly hyperbolic problems to their role in a model NSH system with a coincidence-line flux. The results (summarized in the title of the paper) are unexpected. We conjecture that NSH coincidence-line models may provide good prototypes for mathematical properties and qualitative behavior of a larger class of problems, including some models for porous media flow.

Acknowledgements. The research in this paper was supported in part by the Texas Advanced Research Program under Grant 00365-2124 ARP; the Department of Energy, grant DE-FG05-91ER 25102, and NSF grant DMS-91-03560, with support from the Air Force Office of Scientific Research.

REFERENCES

[1] Allen M. B., Behie G. A., Trangenstein J. A. *Multiphase flow in porous media: mechanics, mathematics, and numerics.* Lecture Notes in Engineering 34. Springer-Verlag, New York, 1988.

[2] Bick J. H., Newell G. F. A continuum model for two-directional traffic flow. *Quart. Appl. Math.*, XVIII:191–204, 1960.

[3] Holden H., Holden L. On the Riemann problem for a prototype of a mixed type conservation law, II. In *Current Progress in Hyperbolic Systems: Riemann Problems and Computations*, Contemporary Mathematics 100, pages 331–367. Amer. Math. Soc., Providence, 1989. B. Lindquist, ed.

[4] Holden H., Holden L., Risebro N. H. Some qualitative properties of 2×2 systems of conservation laws of mixed type. In *Nonlinear Evolution Equations that Change Type*, IMA Volumes in Mathematics and its Applications 27, pages 67–78. Springer, 1990. B. Keyfitz and M. Shearer, eds.

[5] Johansen T., Winther R. The solution of the Riemann problem for a hyperbolic system of conservation laws modelling polymer flooding. *SIAM J. Math. An.*, 19:541–566, 1988.

[6] Keyfitz B. L. Some elementary connections among nonstrictly hyperbolic conservation laws. In *Nonstrictly Hyperbolic Conservation Laws*, Contemporary Mathematics 60, pages 67–77. Amer. Math. Soc., Providence, 1987. B. Keyfitz and H. Kranzer, eds.

[7] Keyfitz B. L. A criterion for certain wave structures in systems that change type. In *Current Progress in Hyperbolic Systems: Riemann Problems and Computations*, Contemporary Mathematics 100, pages 203–213. Amer. Math. Soc., Providence, 1989. B. Lindquist, ed.

[8] Keyfitz B. L. Shocks near the sonic line: a comparison between steady and unsteady models for change of type. In *Nonlinear Evolution Equations that Change Type*, IMA Volumes in Mathematics and its Applications 27, pages 89–106. Springer, 1990. B. Keyfitz and M. Shearer, eds.

[9] Keyfitz B. L. Change of type in simple models of two-phase flow. In *Viscous Profiles and Numerical Approximation of Shock Waves*, pages 84–104. SIAM, Philadelphia, 1991. M. Shearer, ed.

[10] Keyfitz B. L. Conservation laws that change type and porous medium flow: a review. In *Modeling and Analysis of Diffusive and Advective Processes in Geosciences*, pages 122–145. SIAM, Philadelphia, 1992. W. E. Fitzgibbon and M. F. Wheeler, eds.

[11] Keyfitz B. L. A method for generating nonstrictly hyperbolic fluxes with eigenvalue coincidence along a line. In preparation.

[12] Liu T.-P. The Riemann problem for general 2×2 conservation laws. *Amer. Math. Soc. Trans.*, 199:89–112, 1974.

[13] Osher S. J. Riemann solvers, the entropy condition, and difference approximations. *SIAM Jour. Numer. Anal.*, 21:217–235, 1984.

[14] Pego R. L., Serre D. Instabilities in Glimm's scheme for two systems of mixed type. *SIAM Jour. Numer. Anal.*, 25:965–988, 1988.

[15] Rhee H.-K., Aris R., Amundson N. R. *First-Order Partial Differential Equations: Volume I, Theory and Application of Single Equations.* Prentice-Hall, Englewood Cliffs, 1986.

[16] Schaeffer D. G., Shearer M. The classification of 2×2 systems of non-strictly hyperbolic conservation laws, with application to oil recovery. *Comm. Pure Appl. Math.*, 40:141–178, 1987.

[17] Shearer M., Trangenstein J. A. Loss of real characteristics for models of three-phase flow in a porous medium. *Transport in Porous Media*, 4:499–525, 1989.

[18] Stewart H. B., Wendroff B. Two-phase flow: models and methods. *Jour. Comp. Physics*, 56:363–409, 1984.

[19] Temple J. B. The L^1-norm distinguishes the strictly hyperbolic from a nonstrictly hyperbolic theory of the initial value problem for systems of conservation laws. In *Nonlinear Hyperbolic Equations – Theory, Computational Methods and Applications*, Notes Numer. Fluid Mech. 24, pages 608–616. Aachen, 1988; Vieweg, Braunschweig, 1989.

[20] Temple J. B. Systems of conservation laws with coinciding shock and rarefaction curves. In *Nonlinear Partial Differential Equations*, Contemporary Mathematics 17, pages 143–151. American Mathematical Society, Providence, 1983. J. A. Smoller, ed.

[21] Vinod V. *Structural stability of Riemann solutions for a multiphase kinematic conservation law model that changes type.* PhD thesis, University of Houston, 1992.

International Series of Numerical Mathematics, Vol. 114, © 1993 Birkhäuser Verlag Basel

A Central Limit Theorem for Multiscaled Permeability

S. M. Kozlov*

Abstract. Recent experiments of Noetinger and Jacquin [7] showed high accuracy of the effective three–dimensional permeability formula given by the cube of the average of the third root of local permeability. Here, a model of a locally multiscaled lognormal permeability is proposed for which this formula is asymptotically exact. The model reflects the real situation of many (asymptotically infinite) length scales of heterogeneties.

1 INTRODUCTION

It is well known that geometrical and geological properties of porous media are very complex, and the popular mathematical model describes them as realizations of random functions of position (see [6]). Experience says that there exist many levels of heterogeneties, starting from the pore level (see [1]). In principle, one should begin a mathematical investigation with the Stokes equations for the fluid inside the pores. But, as established in the physical literature (see [6], [1]), the hypothesis of local statistical homogeneity of the pore geometry leads directly to the Darcy equations. The homogenization of the Stokes equations in a random domain will be analyzed in a forthcoming paper. Here, it is assumed that local permeability varies due to the presence of larger–scaled heterogeneities. The simplest conjecture is to assume permeability to be a random, statistically–homogeneous field. This assumption leads to the possibility of applying the homogenization theory of random partial differential equations with statistically–homogeneous coefficients to the problems of porous media (see [2] for a review of basic notions of that theory).

To find the effective permeability according to that theory, one should solve an auxiliary equation on a probability space. We shall assume the local permeability tensor to be isotropic, with the local permeability being lognormal of the form

$$K(x) = K_0 \exp\{\lambda F(x)\}, \tag{1.1}$$

where $F(x)$ is a Gaussian field, λ is a dimensionless parameter, and K_0 a normalizing parameter. The model (1.1) is known to be practically reasonable, as indicated by the experiments in Noetinger and Jacquin [7]. In two dimensions, homogenization leads to the explicit formula (see [2])

$$K_{eff} = K_0 \exp\{\lambda\langle F\rangle\}, \tag{1.2}$$

where $\langle F\rangle$ is the mean value of the field F. However, in three dimensions, the asymptotic behavior as $\lambda \to \infty$ is given by

$$K_{eff} \sim K_0\lambda^{\alpha} \exp\{\lambda(\langle F\rangle + F_*\sigma)\}, \tag{1.3}$$

*Laboratoire APT, URA CNRS 225, Université de Provence Aix Marseille I, 3, place V.Hugo, 13331 Marseille France CEDEX

where α is some exponent, F_* is the first channel percolation level from above of the normalized field $(F - \langle F \rangle)/\sigma$, and σ the covariance of F (see [4] for the proof of the discrete version of (1.3)). We also note that (1.3) is consistent with (1.2) since, in two dimensions, $F_* = 0$ and the asymptotics give the exact relation. In two dimensions, the effective permeability formula agrees with the Landau conjecture (see [5]), but the asymptotic formula (1.3) contradicts it. For the constant F_*, only numerical estimates are known (see, for example, [9], and works quoted there), and it seems improbable to find an explicit evaluation for general correlations. Attempts to calculate the effective permeability explicitly are almost hopeless since, according to (1.3), the effective permeability formula also involves the value of the percolation threshold.

However, recent laboratory experiments performed by Noefinger and Jacquin [7] on three-dimensional porous materials such as sandstone and limestone show an excellent agreement between the measured overall permeability and the simple algebraic formula

$$K_{eff} = \left\langle K^{1/3}(x) \right\rangle^3 = \left(\frac{1}{|V|} \sum_i K_i^{1/3} |V_i| \right)^3, \tag{1.4}$$

where K_{eff} and K_i are measured values of the permeabilities of the entire sample V and the small cubes V_i into which it had been decomposed.

According to the considerations above, in order to construct a reasonable theory to explain (1.4), we should reformulate the basic assumption of random homogenization theory, i.e., statistical homogeneity, and take into account the existence of many length scales of heterogeneities. Here, the following model of the local permeability depending on the parameter N (number of length scales) is proposed:

$$K_N(x) = K_0 \exp\left\{ \frac{1}{\sqrt{N}} F_N(x) \right\}, \qquad F_N(x) = \sum_{k=1}^{N} f_k(x), \tag{1.5}$$

where the independent, statistically-homogeneous fields $f_k(x)$ are assumed to have correlation lengths $\xi_k \to \infty$, $k \to \infty$ in such a way that the ratio ξ_{k+1}/ξ_k also converges to infinity. Then, according to the Central Limit Theorem, $\frac{1}{\sqrt{N}} F_N(x)$ converges in distribution for any point x to a normal distribution as $N \to \infty$. So, for large N, the local permeability (1.5) could be considered as lognormal. The aim of this paper is to show that (1.4) holds asymptotically for the local permeability (1.5) as $N \to \infty$.

2 PRELIMINARY RESULTS FROM RANDOM HOMOGENIZATION THEORY

2.1 General notions

Let us first introduce usual notations of random homogenization. Let $(\Omega, \mathcal{F}, \mu)$ be a probability space, where $\mathcal{F}$ is a complete σ-algebra and μ is a probability measure. Assume that the D-dimensional dynamic system $T(x) : \Omega \to \Omega$, $x \in R^D$, acts on Ω and satisfies the following conditions:

1. $T(x) \circ T(y) = T(x + y)$.
2. $T(0) = I$, the identity transformation.

3. The transformation $(x, \omega) \to T(x)\omega$ of $R^D \times \Omega$ is measurable when R^D is endowed with the Lebesgue algebra of sets.

For a system $T(x)$ as above, one can associate the strongly continuous group of unitary operators $U(x) : L^2(\Omega) \to L^2(\Omega)$ defined by

$$(U(x)f)(\omega) = f(T(x)\omega).$$

Then,

$$\lim_{|x|\to 0} \|U(x)f - f\|_{L^2(\Omega)} = 0, \qquad \forall f \in L^2(\Omega);$$

see [10]. Then, there exist infinitesimal generators (see [10]),

$$\partial_i = \lim_{t\to 0} \frac{U_i(t) - E}{t}, \qquad U_i(t) = U((0, \dots, t, \dots 0)),$$

($t \in R$ in place of the ith coordinate), where E is the unit operator. The vector space $H = [L^2(\Omega)]^D$, where

$$f \in H \Leftrightarrow f = (f_1, \dots, f_D), \qquad \|f\|_H^2 = \int_\Omega |f|^2 \mu(d\omega),$$

can be represented as the orthogonal sum (Weyl's decomposition),

$$H = H_p + H_s + R^D,$$

where

$$H_p = \overline{\left\{(\partial_1 f, \dots, \partial_D f); f \in L^2(\Omega) \cap D(\partial_i),\ \forall i\right\}},$$

and the bar means closure in H and $D(\partial_i)$ is the domain of ∂_i. The subspace H_s is defined as ([10])

$$H_s = \left\{f \in H, \langle f \rangle = 0, \int_{R^D} f(T(x)\omega) \cdot \nabla\varphi dx = 0,\ \mu\text{-a.s. } \forall \varphi \in C_0^\infty(R^D)\right\}.$$

Now, let us fix a random, symmetric matrix $a(\omega) = (a_{ij}(\omega))$ such that

$$\delta E \le a \le \delta^{-1} E \quad \mu\text{-a.s.}, \tag{2.1}$$

where $\delta > 0$. To the random matrix a we associate a constant matrix $\hat{a}$ defined by

$$\hat{a}\xi \cdot \xi = \inf_{v \in H_p} \int a(\xi + v) \cdot (\xi + v) d\mu, \quad \forall \xi \in R^D, \tag{2.2}$$

where $\cdot$ stands for the scalar product and a matrix $\hat{b}$ given by

$$\hat{b}\xi \cdot \xi = \inf_{p \in H_s} \int_\Omega a^{-1}(p + \xi) \cdot (p + \xi) d\mu, \quad \forall \xi \in R^D. \tag{2.3}$$

By (2.1) and the standard theory of bounded quadratic forms in Hilbert spaces, we have existence and uniqueness of the solutions to (2.2) and (2.3). Then, by Weyl's decomposition result, we get the natural relation

$$\hat{a} = \hat{b}^{-1}.$$

Choosing $v = 0$ and $p = 0$ in (2.2) and (2.3) leads to the bilateral bound

$$\langle a^{-1} \rangle^{-1} \le \hat{a} \le \langle a \rangle.$$

2.2 Random homogenization theorem

Typically ([4], [10]), homogenization theory is concerned with a random operator of the form

$$A_\varepsilon = \frac{\partial}{\partial x_i}\left(a_{ij}\left(T\left(\frac{x}{\varepsilon}\right)\omega\right)\frac{\partial}{\partial x_j}\right).$$

Theorem 2.1 *Assume that $A_\varepsilon u_\varepsilon = f$ for $f \in H^{-1}(V)$ in the weak sense in a domain $V \subset R^D$ and*

$$u_\varepsilon \rightharpoonup u \quad in \quad H^1_{loc}(V) \ weakly.$$

Then,

$$\hat{A}u = f,$$

where

$$\hat{A} = \sum_{i,j=1}^{D} \frac{\partial}{\partial x_i}\left(\hat{a}_{ij}\frac{\partial}{\partial x_j}\right).$$

To simplify further calculations, we need the cohomological formulation of homogenization. Those readers not familiar with differential geometry notations can use ordinary calculus; this simply leads to longer calculations in places.

Let $*$, $\wedge$, d, δ be the usual notations in differential geometry (see [8]). Introduce k-differential forms on Ω using the standard procedure; their realizations are k-differential forms on R^n. Denote the closure of that space in the $L^2(\Omega)$-topology for the coefficients by $\Lambda^k\Omega$. Then, with the help of the operators ∂_i, we can define the differential de Rham complex:

$$0 \xrightarrow{d} \Lambda^0\Omega \xrightarrow{d} \Lambda^1\Omega \xrightarrow{d} \dots \xrightarrow{d} \Lambda^{D-1}\Omega \xrightarrow{d} \Lambda^D\Omega \xrightarrow{d} 0.$$

We need the following cohomology spaces:

$$H^1 = Ker\{d : \Lambda^1\Omega \to \Lambda^2\Omega\}/Im\{d : \Lambda^0\Omega \to \Lambda^1\Omega\} \cong R^D,$$

$$H^{D-1} = Ker\{d : \Lambda^{D-1}\Omega \to \Lambda^D\Omega\}/Im\{d : \Lambda^{D-2}\Omega \to \Lambda^{D-1}\Omega\} = R^D.$$

To be specific, assume that $a_{ij}(\omega) = a(\omega)\delta_{ij}$, δ_{ij} being the Kroneker delta, and that the fields $a(T(Ux)\omega)$ are statistically equivalent for all cube-preserving linear mappings U. Later, we call the last assumption the *cubic isotropy* of a random field. Under these assumptions, the homogenized matrix is also isotropic and has the following representations:

$$\hat{a}|\xi|^2 = \inf_{[\omega]=\xi}\langle a\omega \wedge *\omega\rangle, \qquad \xi \in H^1, \tag{2.4}$$

and

$$\hat{a}^{-1}|\zeta|^2 = \inf_{[\alpha]=\zeta}\langle a^{-1}\alpha \wedge *\alpha\rangle, \qquad \zeta \in H^{D-1}, \tag{2.5}$$

where $\omega \in \Lambda^1\Omega$ in (2.4) and $d\omega = 0$ in (2.5); $\alpha \in \Lambda^{D-1}\Omega$, $d\alpha = 0$ and $[\cdot]$ is the cohomological class. The spaces H^1 and H^{D-1}, endowed with the euclidian norm, are induced from the original R^D. If we suppose that $\zeta = *\xi$, then the solutions of (2.4) and (2.5) are related by

$$\alpha = *a\omega/\hat{a}.$$

2.3 Regular perturbation formulae

Consider the random matrix $a_{ij}(\omega) = a(\omega)\delta_{ij}$, and assume that $a(\omega) = e^{\gamma f(\omega)}$, where γ is a small parameter and $f(\omega)$ is a bounded function such that

$$\langle f \rangle = 0, \tag{2.6a}$$

$$|f| \leq c_f < \infty, \quad \mu \text{ a.s.}; \tag{2.6b}$$

moreover, assume f to be a cubic isotropy. Denote by $\hat{a}_\gamma$ the homogenized coefficient from (2.2). First-order approximation to the Euler equations for (2.2) leads to

$$\begin{aligned} \Delta\chi_j &= -\frac{\partial}{\partial x_j} f, & \Delta &= \sum_{i=1}^{D} \partial_i^2, \\ v &= \nabla\Big[\gamma\chi \cdot \xi\Big] + \mathcal{O}(\gamma^2), & \chi \cdot \xi &= \sum_{j=1}^{D} \chi_j \xi_j. \end{aligned} \tag{2.7}$$

Introduce the spectral density of the field f:

$$S(\xi) = (2\pi)^{-D} \int_{R^D} e^{ix\xi} \Big\langle f(\omega) f(T(x)\omega) \Big\rangle dx.$$

Then, under the assumption that

$$\int_{R^D} |\xi|^{-2} S(\xi) d\xi \leq c_f < \infty,$$

the solution to (2.7) exists in $L^2(\Omega) \cap D(\partial_i)$, $i = 1, \dots, D$ (see [3]).

Proposition 2.1 *Under the above assumptions,*

$$\hat{a}_\gamma \leq \Big(1 + \Big(\frac{1}{2} - \frac{1}{D}\Big)\langle f^2 \rangle \gamma^2 + c_0 \gamma^3\Big), \tag{2.8}$$

where c_0 depends only on the constant c_f in (2.6b).

Proof: Substitute $v_1 = \gamma \nabla\chi \cdot \xi$ into (2.2). Then,

$$\begin{aligned} \hat{a}_\gamma |\xi|^2 &\leq \Big\langle e\gamma f |\xi + \gamma\nabla\chi \cdot \xi|^2 \Big\rangle \\ &\leq \Big\langle \Big(1 + \gamma f + \frac{1}{2}\gamma^2 f^2\Big) |\xi + \gamma\nabla\chi \cdot \xi|^2 \Big\rangle + c\gamma^3 |\xi|^2 \\ &= \Big(1 + \frac{1}{2}\gamma^2 \langle f^2 \rangle\Big) |\xi|^2 + 2\gamma^2 \Big\langle f \nabla\chi \cdot \xi \Big\rangle \cdot \xi + \gamma^2 \Big\langle \nabla\chi \cdot \xi \cdot \nabla\chi \cdot \xi \Big\rangle + c_1 \gamma^3 |\xi|^2. \end{aligned}$$

From the cubic isotropy of f it follows that

$$\Big\langle f \partial_i \Delta^{-1} \partial_j f \Big\rangle = \rho\delta_{ij}, \qquad \sum_{k=1}^{D} \Big\langle \partial_i \Delta^{-1} \partial_k f \partial_k \Delta^{-1} \partial_j f \Big\rangle = \rho\delta_{ij}, \tag{2.9}$$

where $\rho = D^{-1}\langle f^2 \rangle$, and this yields (2.8). ∎

Proposition 2.2 *Under the same assumptions,*

$$\hat{a}_\gamma \geq \left(1 + \left(\frac{1}{2} - \frac{1}{D}\right)\gamma^2 - c_0\gamma^3\right). \tag{2.10}$$

Proof: To obtain (2.10), it suffices to take $\xi = (1, \ldots, 1)$ and use

$$p = \gamma(Df + \partial_1\chi \cdot \xi, \ldots, Df + \partial_D\chi \cdot \xi)$$

in the representation (2.3). ■

In the next section, we shall need a modified test function. Let $\tilde{\psi} = (\tilde{\psi}_{ij})$, where $\tilde{\psi}_{ij} = \gamma \frac{\partial}{\partial_i}\chi_j$. Then, by cubic symmetry,

$$\left\langle a(E + \tilde{\psi})\tilde{\psi}^T \right\rangle = c_\gamma \delta_{ij},$$

where ψ^T is the transposed matrix and E is the unit matrix. Then by (2.9),

$$\left\langle a\tilde{\psi} \right\rangle = -\left\langle a\tilde{\psi}\tilde{\psi}^T \right\rangle = -\frac{\gamma^2}{D},$$

so that $c_\gamma = \mathcal{O}(\gamma^3)$ as $\gamma \to 0$. Now, set $\beta = \dfrac{\gamma^2/D}{\gamma^2/D + c_\gamma} = 1 + \mathcal{O}(\gamma)$ and $\psi = \beta\tilde{\psi}$. Then, by the definition of ψ,

$$\left\langle a(E + \psi)\psi^T \right\rangle = 0. \tag{2.11}$$

Elsewhere, we will also use the notation χ for $\beta\chi$.

2.4 Cohomological form of perturbation

First develop the variational problems (2.4) and (2.5), assuming that $a = \exp\{\gamma f\}$ and that γ is a small parameter, to see that

$$\hat{a}|\xi|^2 = |\xi|^2 + \gamma^2 \inf_{[\omega_1]=0} \left\langle \omega_1 \wedge *\omega_1 + 2f\omega_1 \wedge *\xi \right\rangle + \mathcal{O}(\gamma^4) \tag{2.12}$$

for $\omega_1 \in \Lambda^1\Omega$ and

$$\hat{a}^{-1}|\zeta|^2 = |\zeta|^2 + \gamma^2 \inf_{[\alpha_1]=0} \left\langle \alpha_1 \wedge *\alpha_1 - 2f\alpha_1 \wedge *\zeta \right\rangle + \mathcal{O}(\gamma^4) \tag{2.13}$$

for $\alpha \in \Lambda^{D-1}\Omega$, respectively. The solution of (2.12) depends linearly on the fixed class ξ; thus,

$$\omega_1 = A\xi, \qquad A : \Lambda^1 \to \Lambda^1.$$

The map A is linear and, since ω_1 is exact,

$$Ad(x \cdot \xi) = d(\mathcal{A}x \cdot \xi),$$

and this relation defines an exact one-form $\mathcal{A}x\cdot dx$, where $dx = (dx_1, \ldots, dx_D)$. Denote this form by χ. The solution to the dual problem (2.13) is related to ω_1 by

$$\alpha_1 = *A\zeta = *A*\zeta \quad \text{for} \quad \zeta = *\xi,$$

so that

$$\alpha_1 = A_*\zeta, \qquad A_* : \Lambda^{D-1}\Omega \to \Lambda^{D-1}\Omega, \qquad A_* = *A*.$$

We define $\chi_* = *\chi \in \Lambda^{D-1}\Omega$ in the same way. Multiplying χ and χ_* by β, we find that $\langle a(E+A)\wedge *A\rangle = 0$.

3 A CENTRAL LIMIT THEOREM FOR THE EFFECTIVE PERMEABILITY

Let $\{f_n(\omega)\}_{n=1}^{\infty}$ be a sequence of identically-distributed, independent random variables. Assume all of them to be uniformly bounded and to have mean value zero:

$$-c_f \le f_n(\omega) \le c_f < \infty, \quad \mu\text{-a.s.}, \qquad \langle f_n\rangle = 0. \tag{3.1}$$

Introduce also a sequence of positive numbers $\{\xi_n\}_{n=1}^{\infty}$ such that $\xi_n \to 0$ as $n \to \infty$. Let $f_n(x) = f_n(T(x)\omega)$, and consider the following CLT-sum:

$$F_N(x) = \frac{1}{\sqrt{N}}\sum_{n=1}^{N} f_n\Big(\frac{x}{\xi_n}\Big).$$

Let us denote the homogenized matrix of the random matrix a for convenience by $\langle a\rangle_H$. The following theorem presents the main result of the paper.

Theorem 3.1 *Let* (3.1) *hold, and let*

$$\sum_{n=1}^{\infty} n(\xi_{n+1}/\xi_n)^2 < \infty. \tag{3.2}$$

If $\sigma^2 = \langle f^2\rangle$, *then*

$$\lim_{N\to\infty}\langle e^{F_N(x)}\rangle_H = \exp\Big\{\Big(\frac{1}{2}-\frac{1}{D}\Big)\sigma^2\Big\}.$$

Proof: For convenience, let $K_N = \exp\{F_N(x)\}$. We will estimate the homogenization coefficient $\langle K_N\rangle_H$ from above and below by use of the variational principles (2.2) and (2.5). To find an upper bound, choose the test function in the form $v = \nabla X_N \cdot \xi$, with

$$X_N(x,\omega) = \sum_{k=1}^{N}\xi_k\prod_{j=1}^{k-1}\Big(E + \psi^j\Big(T\Big(\frac{x}{\xi_j}\Big)\omega\Big)\Big)\chi^k\Big(T\Big(\frac{x}{\xi_k}\Big)\omega\Big), \tag{3.3}$$

where $\chi^j = (\chi_1^j, \ldots, \chi_D^j)$ is the solution of (2.7) with f_j instead of f in the right-hand side and $\psi^j = \nabla\chi^j$, normalized so as to satisfy (2.11). In (3.3), $\prod_{j=1}^{0} = E$ and $\prod_{j=1}^{k-1} A_j = A_1\cdot\ldots\cdot A_{k-1}$. By substituting (3.3) into (2.2), we see that

$$\Big\langle K_N\Big\rangle_H \le Tr\Big\langle K_N(E+\nabla X_N)(E+\nabla X_N)^T\Big\rangle, \tag{3.4}$$

where $Tr(A)$ is the trace of the matrix A divided by the dimension D. Let us calculate the derivative

$$\nabla X_N = \sum_{k=1}^{N} \Big(E + \psi^j\Big(T\Big(\frac{x}{\xi_j}\Big)\omega\Big)\Big)\psi^k\Big(T\Big(\frac{x}{\xi_j}\Big)\omega\Big) + w_1 = w_0 + w_1, \tag{3.5}$$

where

$$w_1 = \sum_{k=1}^{N}\sum_{i=1}^{k-1}\frac{\xi_k}{\xi_i}\prod_{j=1}^{i-1}\Big(E + \psi^j\Big(T\Big(\frac{x}{\xi_j}\Big)\omega\Big)\Big)\nabla\psi^i\Big(T\Big(\frac{x}{\xi_i}\Big)\omega\Big)\cdot$$
$$\cdot\prod_{j=i+1}^{k-1}\Big(E + \psi^j\Big(T\Big(\frac{x}{\xi_j}\Big)\omega\Big)\Big)\chi^k\Big(T\Big(\frac{x}{\xi_k}\Big)\omega\Big).$$

Develop the right hand side of (3.4). Let

$$Tr\Big\langle K_N(E + \nabla X_N)(E + \nabla X_N)^T\Big\rangle = T_1 + T_2,$$

where

$$\begin{aligned} T_1 &= Tr\Big\langle K_N(E + \nabla X_N)\Big\rangle, \\ T_2 &= Tr\Big\langle K_N(E + \nabla X_N)\nabla X_N^T\Big\rangle. \end{aligned}$$

Setting $a_i = \exp\{f_i/\sqrt{N}\}$, we have

$$\begin{aligned} T_1 &= Tr\Big\langle \prod_{i=1}^{N} a_i\Big(T\Big(\frac{x}{\xi_i}\Big)\omega\Big)\Big(E + \sum_{k=1}^{N}\prod_{j=1}^{k-1}\Big(E + \psi^j\Big(T\Big(\frac{x}{\xi_j}\Big)\omega\Big)\Big)\psi^k\Big(T\Big(\frac{x}{\xi_k}\Big)\omega\Big)\Big) \\ &\qquad + K_N w_1\Big\rangle \\ &= \langle a\rangle^N - \sum_{k=1}^{N}\langle a\rangle^{N-k}\Big(\langle a\rangle - \langle a\rangle_{H_2}\Big)\langle a\rangle_{H_2}^{k-1} + Tr\langle K_N w_1\rangle \\ &= \langle a\rangle^N - \sum_{k=1}^{N}\langle a\rangle^{N-k+1}\langle a\rangle_{H_2}^{k-1} - \sum_{k=1}^{N}\langle a\rangle^{N-k}\langle a\rangle_{H_2}^{k} + Tr\langle K_N w_1\rangle \\ &= \langle a\rangle_{H_2}^N + Tr\langle K_N w_1\rangle, \end{aligned}$$

where

$$\langle a\rangle_{H_2} = \Big\langle a(E+\psi)(E+\psi)^T\Big\rangle = \Big\langle a(E+\psi)\Big\rangle;$$

the index i was omitted since the a_i are identically distributed. Note that $\langle a_i\nabla\psi^i\rangle = 0$ and $\langle a_i\chi^i\rangle = 0$, since χ^i is an antisymmetric random field while a_i is symmetric. Thus,

$$Tr\langle K_N w\rangle = \sum_{k=1}^{N}\sum_{i=1}^{k-1}\frac{\xi_k}{\xi_i}\langle a_i\psi^i\rangle\langle a_k\chi^k\rangle\langle a\rangle_{H_2}^{k-2} = 0.$$

Finally, we have the relation

$$T_1 = \langle a\rangle_{H_2}^N.$$

Let us evaluate T_2 by the same symmetry-antisymmetry arguments. Split T_2 as follows:

$$T_2 = Tr\Big\langle K_N(E+w_0)w_0^T\Big\rangle + Tr\Big\langle K_N w_1 w_1^T\Big\rangle = T_2^1 + T_2^2. \tag{3.6}$$

By (3.5),

$$\begin{aligned} T_2^1 &= Tr\Big\langle \prod_{i=1}^N a_i\Big(E+\sum_{k_1=1}^N\prod_{j_1=1}^{k_1-1}(E+\psi^{j_1})\psi^{k_1}\Big)\Big(\sum_{k_2=1}^N\prod_{j_2=1}^{k_2-1}(E+\psi^{j_2})\psi^{k_2}\Big)^T\Big\rangle \\ &= Tr\Big\langle \prod_{i=1}^N a_i\sum_{k_2=1}^N\prod_{j_2=1}^{k_2-1}(E+\psi^{j_2})\psi^{k_2}\Big\rangle \\ &\quad + Tr\Big\langle \prod_{i=1}^N a_i\sum_{k_1,k_2=1}^N\prod_{j_1=1}^{k_1-1}(E+\psi^{j_1})\psi^{k_1}\Big(\prod_{j_2=1}^{k_2-1}(E+\psi^{j_2})\psi^{k_2}\Big)^T\Big\rangle \\ &= \langle a\rangle_{H_2}^N - \langle a\rangle^N + Tr\Big\langle \prod_{i=1}^N a_i\sum_{k=1}^N\prod_{j=1}^{k-1}(E+\psi^j)\psi^k\psi^{kT}\Big(\prod_{j=1}^{k-1}(E+\psi^j)\Big)^T\Big\rangle, \end{aligned}$$

as a consequence of the independence of a_i for different i and (2.11). The relations

$$\langle a\rangle_{H_2} = Tr\Big\langle a(E+\psi)(E+\psi)^T\Big\rangle \quad\text{and}\quad \Big\langle a\psi\psi^T\Big\rangle = \langle a\rangle - \langle a\rangle_{H_2}$$

imply that

$$\begin{aligned} T_2^1 &= \langle a\rangle_{H_2}^N - \langle a\rangle^N + \sum_{k=1}^N\langle a\rangle^{N-k}\langle a\rangle_{H_2}^{k-1}(\langle a\rangle - \langle a\rangle_{H_2}) \\ &= \langle a\rangle_{H_2}^N - \langle a\rangle^N + \sum_{k=1}^N\langle a\rangle^{N-k+1}\langle a\rangle_{H_2}^{k-1} - \sum_{k=1}^N\langle a\rangle^{N-k}\langle a\rangle_{H_2}^k \\ &= 0. \end{aligned}$$

Now, consider the second term of (3.6):

$$\begin{aligned} T_2^2 &= Tr\Big\langle K_N w_1 w_1^T\Big\rangle = \sum_{k_1,k_2=1}^N Tr\Big\langle \prod_{i=1}^N a_i\sum_{i_1=1}^{k_1-1}\sum_{i_2=1}^{k_2-1}\frac{\xi_{k_1}}{\xi_{i_1}}\frac{\xi_{k_2}}{\xi_{i_2}}\prod_{j_1=1}^{i_1-1}(E+\psi^{j_1})\nabla\psi^{i_1}\cdot \\ &\quad\cdot\prod_{j_1=i_1+1}^{k_1-1}(E+\psi^{j_1})\chi^{k_1}\Big(\prod_{j_2=1}^{i_2-1}(E+\psi^{j_2})\nabla\psi^{i_2}\prod_{j_2=i_2+1}^{k_2-1}(E+\psi^{j_2})\chi^{k_2}\Big)^T\Big\rangle. \end{aligned}$$

The symmetry–antisymmetry argument implies that all terms with either $i_1 \neq i_2$ or $k_1 \neq k_2$ vanish, and

$$\begin{aligned} T_2^2 &= \sum_{k=1}^N Tr\Big\langle \prod_{i=1}^N a_i\sum_{i=1}^{k-1}\Big(\frac{\xi_k}{\xi_i}\Big)^2\prod_{j_1=1}^{i-1}(E+\psi^{j_1})\nabla\psi^i\prod_{j_1=i+1}^{k-1}(E+\psi^{j_1})\chi^k\cdot \\ &\quad\cdot\Big(\prod_{j_2=1}^{i-1}(E+\psi^{j_2})\nabla\psi^i\prod_{j_2=i+1}^{k-1}(E+\psi^{j_2})\chi^k\Big)^T\Big\rangle \\ &= \sum_{k=1}^N\Big[\langle a\rangle_{H_2}^{k-1}\langle a\chi\cdot\chi\rangle\langle a\nabla\psi:\nabla\psi\rangle\langle a\rangle^{N-k-1}\Big(\sum_{i=1}^{k-1}\Big(\frac{\xi_k}{\xi_i}\Big)^2\Big)\Big] \\ &= \beta_0\beta_2\sum_{k=1}^N\langle a\rangle_{H_2}^{k-1}\langle a\rangle^{N-k-1}\varepsilon_k, \end{aligned}$$

where : denotes the inner product of tensors and

$$\beta_0 = \langle a\chi \cdot \chi\rangle, \quad \beta_2 = Tr\langle a\nabla\psi : \nabla\psi\rangle, \quad \varepsilon_k = \sum_{i=1}^{k-1} (\xi_k/\xi_i)^2.$$

The indices are omitted since the distributions of the random variables do not depend on them. Finally,

$$\langle K_N\rangle_H \leq \langle a\rangle_{H_2}^N + \beta_0\beta_2 \sum_{k=1}^{N} \langle a\rangle_{H_2}^{k-1} \langle a\rangle^{N-k+1} \varepsilon_k.$$

We can produce a sharper estimate if, instead of X_N, we use the test function

$$X_{N_1,N} = \sum_{k=N_1}^{N} \xi_k \prod_{j=N_1}^{k-1} \left(E + \psi^j\left(T\left(\frac{x}{\xi_j}\right)\omega\right)\right)\chi^k\left(T\left(\frac{x}{\xi_k}\right)\omega\right).$$

By the same argument,

$$\begin{aligned} \langle K_N\rangle_H &\leq \langle a\rangle^{N_1}\left[\langle a\rangle_{H_2}^{N-N_1} + \beta_0\beta_2 \sum_{k=N_1}^{N} \langle a\rangle_{H_2}^{k-1} \langle a\rangle^{N-k-1} \varepsilon_{k,N_1}\right], \\ \varepsilon_{k,N_1} &= \sum_{i=N_1}^{k-1} \left(\frac{\xi_k}{\xi_i}\right)^2. \end{aligned} \tag{3.7}$$

By §2,

$$\begin{aligned} \langle a\rangle_{H_2} &= 1 + \left(\frac{1}{2} - \frac{1}{D}\right)\frac{\sigma^2}{N} + \mathcal{O}(N^{-2}), \\ \langle a\rangle &= 1 + \frac{\sigma^2}{N} + \mathcal{O}(N^{-2}), \end{aligned}$$

so that passing to the limit in (3.7) as $N \to \infty$ gives

$$\begin{aligned} \langle K_N\rangle_H &\leq \exp\left\{\left(\frac{1}{2} - \frac{1}{D}\right)\sigma^2\right\} + \beta_0\beta_2 e^{\sigma^2} \sum_{k=N_1}^{\infty} \varepsilon_{k,N_1} \\ &\leq \exp\left\{\left(\frac{1}{2} - \frac{1}{D}\right)\sigma^2\right\} + \sum_{k=N_1}^{\infty} k(\xi_k/\xi_{k-1})^2. \end{aligned}$$

Then, (3.2) shows that, as $N_1 \to \infty$,

$$\limsup_{N\to\infty} \langle K_N\rangle_H \leq \exp\left\{\left(\frac{1}{2} - \frac{1}{D}\right)\sigma^2\right\},$$

which is the desired upper bound.

In order to get the lower bound, we use the representation (2.5) and its approximation (2.13). Then, take the test $(D-1)$-form in (2.5) to be

$$Y_{N,N_1} = \sum_{k=N_1}^{N} \xi_k \prod_{j=N_1}^{k-1} A_*^j\left(T\left(\frac{x}{\xi_j}\right)\omega\right)\chi_*^k\left(T\left(\frac{x}{\xi_k}\right),\omega\right),$$

where $\prod$ is now the composition of the linear maps on $\Lambda^{D-1}\Omega$. Then, analogous calculations to those above lead to the estimate

$$\langle K_N\rangle_H^{-1} \leq \langle a^{-1}\rangle^{N_1}\Big(\langle a\rangle_{H_2}^{N-N_1} + \beta_0\beta_2 \sum_{k=N_1}^{N} \langle a\rangle_{H_2}^{1-k}\langle a^{-1}\rangle^{N-k-1}\varepsilon_{k,N_1}\Big),$$

and the theorem follows. ■

Remark. The same limit holds if we replace the assumptions on $\{\xi_k\}$ by the analogous assumptions that $\xi_k \to \infty$ as $k \to \infty$ and that

$$\sum_{k=1}^{\infty}\left(\frac{\xi_k}{\xi_{k+1}}\right)^2 k < \infty.$$

The trial function in (2.4) could then be taken in the form

$$X_{N,N_1} = \sum_{k=N_1}^{N} \frac{\xi_k}{\xi_N} \prod_{j=N_1}^{k-1} A^j\Big(T\Big(\frac{\xi_N}{\xi_j}\Big)\omega\Big)\chi^k\Big(T\Big(\frac{x\xi_N}{\xi_k}\Big)\omega\Big).$$

Here, the scaling invariance of the homogenized coefficient was used:

$$\Big\langle K_N(x)\Big\rangle_H = \Big\langle K_N(x\xi_N)\Big\rangle_H.$$

REFERENCES

[1] Dagan G. *Flow and Transport in Porous Formations.* Springer-Verlag, Berlin, New York, 1989.

[2] Kozlov S. M. Averaging of random operators. *Matem. Sbornik*, 109(2):188–203, 1979.

[3] Kozlov S. M. The method of averaging and random walks in inhomogeneous environments. *Russian Math. Surveys*, 40(2):73–145, 1985.

[4] Kozlov S. M. Geometric aspects of homogenization. *Russian Math. Surveys*, 44(2):91–144, 1989.

[5] Landau L. D., Lifshitz E. M. *Electrodynamics of Continuous Media.* Course of Theoretical Physics 8. Pergamon Press, New York, 1984.

[6] Matheron G. *Éléments pour une Theorie des Milieux Poreux.* Masson, Paris, 1967.

[7] Noetinger B., Jacquin C. Experimental tests of a simple permeability composition formula. Society of Petroleum Engineers Preprint SPE 22841, 1991.

[8] Novikov S. P., Dubrovin B. A., Fomenko A. T. *Modern Geometry.* Nauka, Moscow, 1979. In Russian.

[9] Stauffer D. *Introduction to Percolation Theory.* Taylor and Francis Ltd., London, 1985.

[10] Zhykov V. V., Kozlov S. M., Oleinik O. A. *Homogenization of Differential Operators.* Springer-Verlag, Berlin, New York, 1993.

International Series of Numerical Mathematics, Vol. 114,

Front Tracking for the Unstable Hele-Shaw and Muskat Problems

Gunter H. Meyer*

Abstract. A line iterative method is used to solve the free boundary problem of immiscible displacement of a heavy fluid by a light fluid. The influence of surface tension on the stabilization of fluid interfaces is demonstrated.

1 INTRODUCTION

On a macroscopic scale, the flow of a fluid through a porous medium is generally modeled with Darcy's law which at first approximation leads to a diffusion equation for the pressure distribution in the liquid. However, for certain parameter ranges the resulting boundary value problems may be ill-posed, and thus represent sizeable challenges for their numerical solution. We shall consider two such problems of long standing, namely the Hele-Shaw suction problem and the Muskat problem for the immiscible displacement of a viscous fluid by another of lower viscosity. We shall demonstrate that once the problem is regularized by the addition of surface tension then simple front tracking can be employed to follow the evolution of the free boundaries, at least as long as their topology remains simple.

This report consists of two parts. First we outline the results of some numerical experiments showing the effect of surface tension on the stabilization of otherwise unstable free boundaries. The second part summarizes the approach taken to track a free boundary whose mathematical characterization involves its curvature.

2 THE MODELS AND NUMERICAL RESULTS

2.1 The Hele-Shaw problem

The fluid pressure in a sheet of fluid entrapped between two closely-spaced parallel plates is modeled by the following free boundary problem:

$$\Delta u = 0 \quad \text{in } D \tag{2.1}$$

where u is the fluid pressure in a two-dimensional domain D, and where the unknown and moving free boundary ∂D of the fluid is at ambient pressure (taken as zero)

$$u = 0, \tag{2.2}$$

and moves with the Darcy velocity

$$\frac{\partial u}{\partial n} = -kv. \tag{2.3}$$

*School of Mathematics, Georgia Institute of Technology, Atlanta, GA 30332 USA, meyer@math.gatech.edu

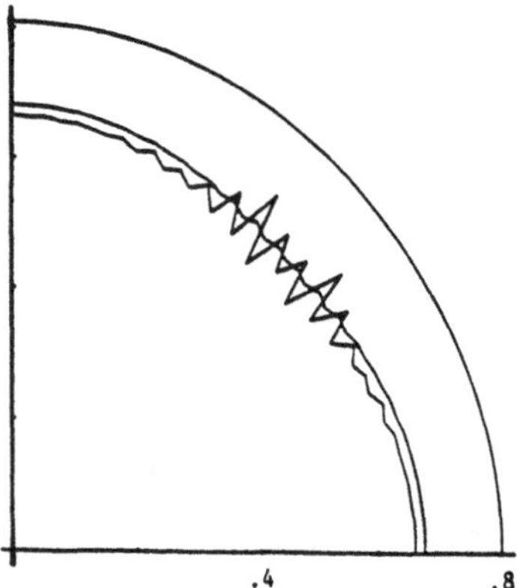

Figure 1: Numerical Instability of the Free Boundary for a Contracting Circular Reservoir.

Here v is the speed of ∂D in the normal direction. We remark that this problem is formally identical with the one-phase Stefan problem when sensible heat is negligible.

If at some point within the fluid, taken as the origin, more fluid is added, then the sheet grows. The resulting free boundary problem is quite stable and has been discussed a great deal in the mathematical and numerical analysis literature (see, e.g., the discussion in Elliott and Ockendon [2]). If, on the other hand, fluid is sucked out at the origin, then only a circular initial reservoir will allow total extraction of the fluid. Any other shape will lead to cusping of the free boundary and attendant breakdown of the model. Hence the problem is unstable because we have no continuity with respect to the initial configuration (see Howison *et al.* [3]). Nonetheless, numerical studies have been published which showed that the free boundary could be tracked almost to the time of cusping and that the numerical results followed the analytical free boundary in certain model cases (see Aitchison [1], Meyer [7]). However, in analogy to the instability of the planar interface for an undercooled Stefan problem (Strain [11]), we should likewise expect that all perturbations of the free boundary of the Hele-Shaw cell will grow unboundedly. Hence, we contend that the earlier numerical studies of the Hele-Shaw problem with suction but no surface tension, although qualitatively correct initially, have no predictive value because the physical instability of the problem is mirrored by numerical instability. For any given set of space and time mesh parameters, front tracking will follow the free boundary until the numerical results begin to oscillate at the smallest spatial length scale of the computation. Figure 1 is a typical result when one tries to follow a contracting circular reservoir with a two-dimensional front-tracking code. Perturbations of the numerical results are the consequence of round-off and truncation errors. The time of appearance for these oscillations depends on the number of computations and hence on the time steps and mesh parameters in the algorithm described below. The location of the free boundary just prior to the appearance of oscillations, both for no and positive surface tension σ, is consistent with the numerical solution of the explicit differential equation for the free boundary

$$\frac{ds}{dt} = -\frac{(\sigma/s - \ln \gamma_0)}{s \ln s/\gamma_0}, \qquad s(0) = s_0,$$

which for $\sigma = 0$ has an analytic solution $t = t(s)$.

The effect of surface tension on the Hele-Shaw cell is examined in some detail by Howison *et al.* [3]. The ambient pressure on the free boundary is now replaced by the capillarity condition

$$u = \sigma/\rho,$$

where σ is proportional to the surface tension of the fluid and ρ is the radius of curvature of the free boundary. Simple qualitative maximum principle-based arguments indicate that pressure gradients are reduced at the tips on the free boundary bulging toward the origin, and increased at the tips of bulges pointing away from the origin which slows and speeds up the motion, respectively. As in the Stefan problem with a Gibbs-Thomson interface condition, we expect a fine balance between the instability of the fluid motion and the smoothing effect of surface tension.

Figures 2-4 show the result of front tracking for a free boundary which at the initial time is the sinusoidally perturbed circle

$$s(\theta, t) = .8 + .1\cos(24\theta).$$

The free boundary moves due to the sink

$$u(r_0, \theta, t) = \log r_0, \qquad r_0 \ll 1.$$

The effect of σ on the smoothing of the free surface is apparent. We are not aware of a theoretical prediction of the damping or magnification of sinusoidal perturbations based on an asymptotic stability analysis of (2.1–2.3) in polar coordinates. However, it may be noted from Figure 4 that a much longer wave-length perturbation establishes itself as time progresses. This behavior is qualitatively consistent with the theoretical (Strain [11]) and numerical (Meyer and Singleton [9]) results for a sinusoidally perturbed traveling wave solution for the Gibbs-Thomson problem.

At this stage our numerical code is not able to check whether the channel will actually develop which is predicted (heuristically) in Howison *et al.* [3] for $\sigma > 0$ as replacement for a cusp which would occur when $\sigma = 0$. The code fails because the preassigned tracking direction and the free boundary become parallel.

2.2 The Muskat problem

We assume that in a channel a plug of fluid with mobility k_2 displaces a second fluid with mobility k_1. Motion occurs due to a pressure differential between the channel inlet and outlet. The mathematical model is

$$\begin{aligned} k_1\Delta u_1 - u_{1_t} &= 0, && 0 < z < s(x,t), \quad x \in (0,1), \\ k_2\Delta u_2 - u_{2_t} &= 0, && s(x,t) < z < 1, \\ \frac{\partial u_i}{\partial x} &= 0, && \text{on } x = 0, 1. \end{aligned}$$

On the interface $z = s(x,t)$, a capillarity pressure balance exists

$$u_2 - u_1 = \sigma/\rho.$$

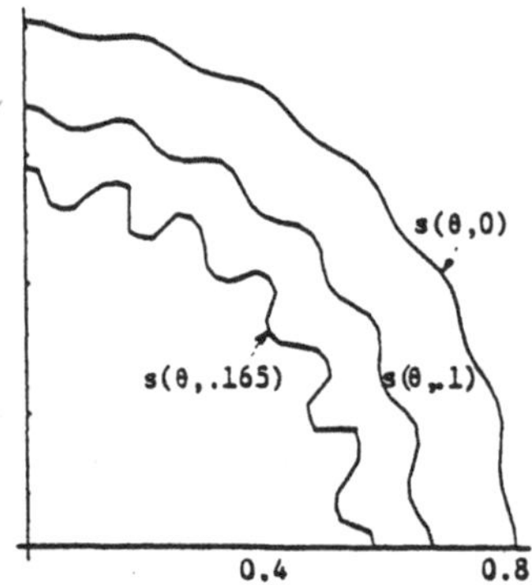

Figure 2: Evolution of the Free Boundary for $\sigma = .0013$. Front Tracking Fails at $t = .165$.

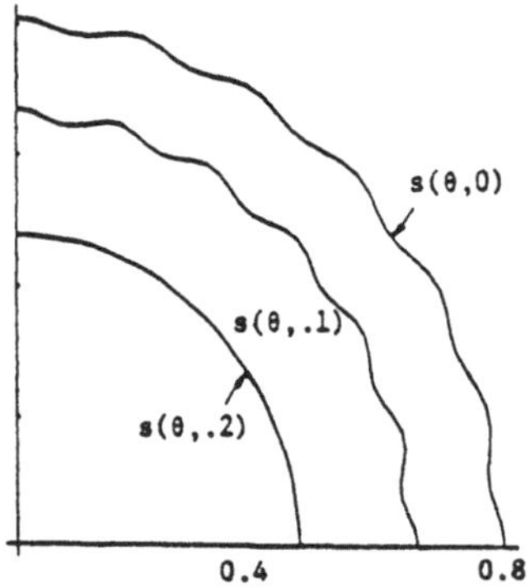

Figure 3: Evolution of the Free Boundary for $\sigma = .0014$. Front Tracking Reaches the Specified Final Time $t = .2$.

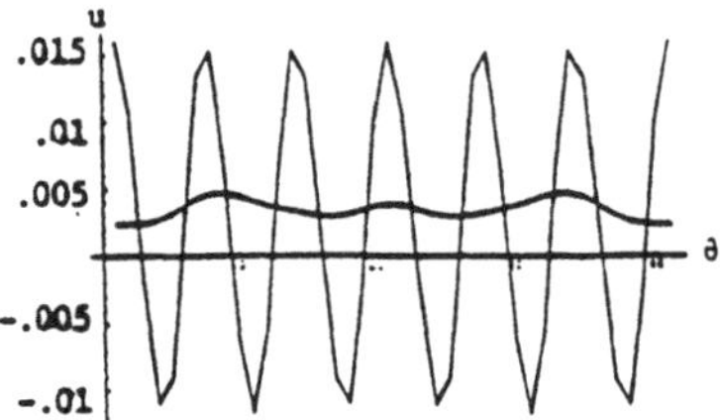

Figure 4: Initial and Final Pressure Distribution on the Free Boundary for $\sigma = .0015$. A Longer Wave Length Perturbation is Established at $t = .221$.

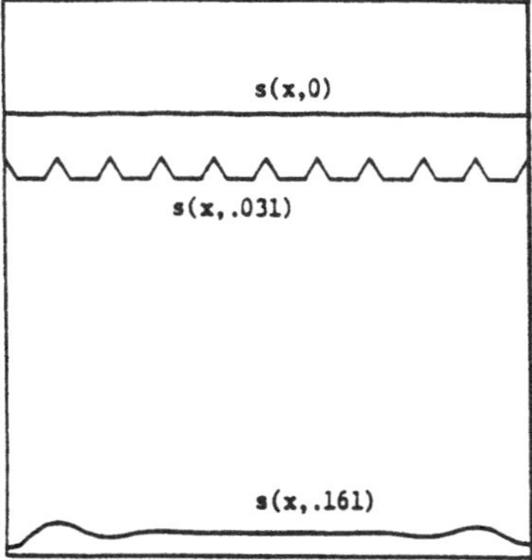

Figure 5: Initial and Final Interface for $\sigma = 0$ at the Time of Failure $t = .031$ and for $\sigma = .001$ at the Final Time $t = .161$.

Continuity of flux and motion of the interface with the Darcy velocity are given by the following two conditions:

$$k_1 \frac{\partial u_1}{\partial n} = k_2 \frac{\partial u_2}{\partial n} = -v, \tag{2.4}$$

where, again, v is the velocity of the free boundary normal to itself. Some comments on the analytic aspects of the Muskat (also known as Verigin) problem may be found in Jiang and Chen [4].

It is well known that if a light fluid displaces a heavy fluid ($k_2 > k_1$) then fingering occurs and for $\sigma = 0$ an initially planar interface becomes completely unstable. Figure 5 illustrates the effect of surface tension on a sinusoidal initial interface

$$s(x,0) = .8 + .001\cos(20\pi x)$$

in a reservoir $[0,1] \times [0,1]$ with impermeable sides which is initially at rest. The permeabilities are $k_1 = 1$ and $k_2 = 4$. The boundary begins to move as the inlet and outlet pressures are lowered and raised with time according to

$$u_1(x,0,t) = -u_2(x,1,t) = -(1 - e^{-100t}). \tag{2.5}$$

For no surface tension and the indicated mesh parameters, the calculation breaks down at time $t_f = .031$. For $\sigma = .001$ we reach the lower boundary at time $t_f = .161$. The damping effect on high frequency oscillations is observable.

Finally, we show in Figure 6 the evolution of the free boundary

$$s(x,0) = .9$$

due to spatially varying boundary pressures obtained by multiplying (2.5) with $\exp(-100(x - .5)^2)$. The non-monotone motion of the free boundary is surprising and seems analogous to the ridging observed for similar boundary data in the Stefan problem with a Gibbs-Thomson interface condition (Meyer and Singleton [9], Strain [11]). However, it is not a curvature effect since the same run with $\sigma = 0$ proceeds smoothly until $t = .06$ and shows the same non-monotone motion of the free boundary.

We remark that in all of these calculations the boundary and initial conditions may be freely chosen because the numerical method is fully implicit and requires no consistency in the data.

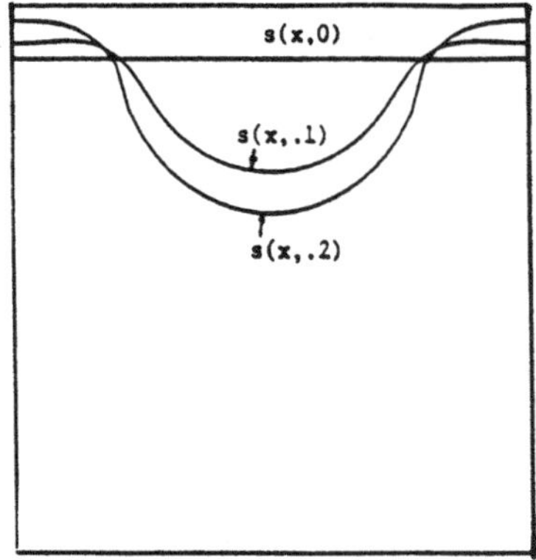

Figure 6: Free Boundary with Non-Monotone Motion Due to Spatially Varying Inlet and Outlet Pressures.

3 NUMERICAL ISSUES

Since the Laplace and constant coefficient heat equations describe the pressure distributions, both free boundary problems can be solved with boundary integral and vortex methods under mild restrictions on the topology of the free boundary. This allows the modeling of complicated fingering patterns (Masukawa and Horne [5], Meiburg and Homsy [6]). Only if the free boundary is initially simple in some convenient coordinate system can our front tracking be employed. On the other hand, our method is well-suited for finite physical domains with general field equations and a variety of interface and boundary conditions. In fact, the above numerical results were obtained with a general purpose free boundary code for nonlinear parabolic systems with variable coefficients (possibly depending on u) which applies equally to the Gibbs-Thomson, Hele-Shaw, and Muskat problems on physical domains in one, two or three space dimensions. The code has been described in detail (Meyer [8]); here, we shall only comment on some of the non-standard considerations peculiar to the Hele-Shaw and Muskat problems.

The motion of the free boundary is tracked along a priori chosen directions, such as the radial direction in the Hele-Shaw problem and the z-direction in the Muskat problem. This imposes the requirement that the boundary be single-valued for fixed t in θ and x. We assume that this condition is true at $t = 0$ and track the boundary only as long as it remains true.

Extensive experience with the Gibbs-Thomson and journal bearing problems have shown that front tracking with curvature boundary conditions succeeds only if the derivatives of the pressure on the interface are restricted to tangential derivatives and those in the tracking direction. Thus, we replace $\dfrac{\partial u}{\partial \theta}$ and $\dfrac{\partial u_i}{\partial x}$ in the interface condition (2.3) and (2.4) with

$$\frac{Du}{D\theta} = \frac{\partial u}{\partial \theta} + \frac{\partial u}{\partial r}\frac{\partial s}{\partial \theta}$$

and

$$\frac{Du_i}{Dx} = \frac{\partial u_i}{\partial x} + \frac{\partial u_i}{\partial z}\frac{\partial s}{\partial x}, \tag{3.1}$$

where $Du/D\theta$ and Du/Dx denote tangential derivatives on $r = s(\theta, t)$ and $z = s(x, t)$. For example, for the Muskat problem the following relations must hold on the free boundary:

$$\frac{\partial s}{\partial t} = -(1 + s_x^2) k_1 \frac{\partial u_1}{\partial z} + s_x k_1 \frac{D u_1}{D x}, \tag{3.2}$$

$$(1 + s_x^2)\left(k_1 \frac{\partial u_1}{\partial z} - k_2 \frac{\partial u_2}{\partial z}\right) - s_x \left(k_1 \frac{D u_1}{D x} - k_2 \frac{D u_2}{D x}\right) = 0. \tag{3.3}$$

Discretization of the problem involves first the replacement of all derivatives with respect to time by backward difference quotients so that in all cases (time implicit) elliptic free boundary problems result. In the Hele-Shaw problem, we then replace all derivatives with respect to θ by central difference quotients, while in the Muskat problem central differences with respect to x are used. The original free boundary problem is now approximated by a system of coupled ordinary differential equations subject to boundary and interface conditions. Again, for the Muskat problem the system

$$\begin{aligned} L_i u_{ij} &\equiv k_i u_{ij}(z)'' - \left[\frac{2k_i}{\Delta x^2} + \frac{1}{\Delta t}\right] u_{ij}(z) \\ &= F_{ij}(z), \qquad i = 1, 2, \quad j = 1, .., M, \\ F_{ij}(z) &= -\frac{k_i [u_{ij+1}(z) + u_{ij-1}(z)]}{\Delta x^2} - \frac{u_{ij,n-1}(z)}{\Delta t}, \end{aligned} \tag{3.4}$$

results. In the interface conditions (2.4) and (3.1) on $z = s_j \cong s(x_j, t)$ we write second- or fourth-order differences. For example,

$$\begin{aligned} s_x &= \frac{s_{j+1} - s_{j-1}}{2\Delta x}, \quad \frac{D u_i}{D x} = \frac{u_{ij+1}(s_{j+1}) - u_{ij-1}(s_{j-1})}{2\Delta x}, \\ s_{xx} &= \frac{s_{j+1} + s_{j-1} - 2 s_j}{\Delta x^2}, \quad \frac{\partial s}{\partial t} = \frac{s_j - s_{j,n-1}}{\Delta t}, \end{aligned}$$

where $u_{ij}(z) \equiv u_{ij,n}(z)$ is the approximation to $u_i(x_j, z, t_n)$ along the line $x = x_j$ at time $t = t_n$ and $s = s_j$ is the free boundary on this line. This system is solved with the line iterative method

$$L_i u_{ij}^k = F_{ij}(z),$$

where for the functions $u_{ij\pm1}(z)$ in $F_{ij}(z)$ and the boundary locations $s_{j\pm1}$ in (3.4) we use the latest available data. Hence, only free boundary problems for a single second-order ordinary differential equation occur. In our code, these scalar problems are handled with the sweep (invariant imbedding) method. Specifically, u_{1j} and u_{2j} along the line $x = x_j$ at $t = t_n$ have the representation

$$u_{ij} = R_{ij}(z) v_{ij}(z) + w_{ij}(z), \qquad v_{ij} = k_i u_{ij}, \tag{3.5}$$

where R_{ij} and w_{ij} satisfy given initial value problems for well-defined ordinary differential equations and hence are (in principle) known functions of z; see Meyer [8]. The representations (3.5) and the interface conditions (3.2) are used as follows to find the free boundary s_j on the line $x = x_j$. At s_j we know that the boundary conditions (3.2) must hold (where derivatives with respect to t and x are replaced by difference quotients). Algebraically, one can eliminate u_{ij} and v_{ij} from these expressions and obtain the relationship

$$\phi_j(s_j) = u_2(s_i) - u_1(s_j) - \sigma/\rho,$$

where

$$u_{ij}(s_j) = R_{ij}(s_j)v_{ij}(s_j) + w_{ij}(s_j),$$

and where $v_{1j}(s_j)$ is found from (3.2) and $v_{2j}(s_j)$ from (3.3). Hence, if we evaluate $\phi_j(z)$ along the line then the location of the free boundary must be a solution of

$$\phi_j(z) = 0.$$

Once s_j is known, the pressure along the line is found from the reverse sweep as described, for example, in Meyer [8]. If ϕ has multiple roots, then we choose the s_j which is consistent with a smooth interface $s(x,t)$ in time and space. In practice, this determination is straightforward. The run is terminated when no solution of $\phi_j(z) = 0$ can be found.

All equations were solved numerically on a Sparcstation Ipx or an Iris Indigo without need for any external program libraries as outlined in Meyer [8]. It is, however, difficult to integrate the Riccati differential equation in the Hele-Shaw problem near the well radius $r_0 = .001$ because of large coefficients. This problem was overcome by combining front tracking with a Schwarz alternating procedure on overlapping domains $D_1 = \{(r,\theta) : r_0 < r < .62\}$ and $D_2 = \{(r,\theta) : .4 < r < s(\theta,t), 0 < \theta < \pi/2\}$, and employing a separation of variables solution on D and front tracking on D. A similar technique was used in Ohring [10] to speed up the computation of waves in a sloshing tank by applying a fast Poisson solver in a rectangle below the free boundary and front tracking in an active layer above the rectangle.

The Hele-Shaw computation shown above was carried out for the following parameters, with symmetry being specifically exploited:

$$\Delta\theta = \pi/80, \qquad \Delta r = .001, \qquad \Delta t = .005.$$

The results for the Muskat problem were obtained with the following parameters:

$$\Delta x = .025, \qquad \Delta y = .05, \qquad \Delta t = .005.$$

In all cases, the performance of the method improved dramatically by using fourth-order central difference approximations for all derivatives on the free boundary.

Acknowledgement. This research was supported in part by the DFG under a grant to the Technical University of Munich, Germany.

REFERENCES

[1] Aitchison J. M. Computation of Hele-Shaw flows with free boundaries. *J. Comp. Phys.*, 60:376–390, 1985.

[2] Elliott C. M., Ockendon J. R. *Weak Variational Methods for Moving Boundary Problems.* Pitman Research Notes in Mathematics 59. Longman Scientific and Technical, Harlow, Essex, UK, 1982.

[3] Howison S. D., Lacey A. A., Ockendon J. R. Hele-Shaw free boundary problems with suction. *Q. J. Mech. Appl. Math.*, 41:183–192, 1988.

[4] Jiang L., Chen Z. Weak formulation of a multi-dimensional Muskat problem. In *Free Boundary Problems: Theory Applications. Volume II*, Pitman Research Notes in Mathematics 186. Longman Scientific and Technical, Harlow, Essex, UK, 1990. K.-H. Hoffmann, J. Sprekels, eds.

[5] Masukawa J., Horne R. N. Application of the boundary integral method to immiscible displacement problems. *SPE Res. Engng.*, 3:1069–1077, 1988.

[6] Meiburg E., Homsy G. M. Nonlinear unstable viscous fingers in Hele-Shaw flows. II Numerical simulation. *Phys. Fluids*, 31:429–439, 1988.

[7] Meyer G. H. Hele-Shaw flow with a cusping free boundary. *J. Comp. Phys.*, 44:262–276, 1981.

[8] Meyer G. H. On computing free boundaries which are not level sets. In *Free Boundary Problems: Theory Applications. Volume I*, Pitman Research Notes in Mathematics 185. Longman Scientific and Technical, Harlow, Essex, UK, 1990. K.-H. Hoffmann, J. Sprekels, eds.

[9] Meyer G. H., Singleton D. B. Front tracking for the supercooled Stefan problem. To appear in *Surveys on Mathematics for Industry.*

[10] Ohring S. The numerical computation of nonlinear water waves. *J. Comp. Phys.*, 39:137–163, 1981.

[11] Strain J. A boundary integral approach to unstable solidification. *J. Comp. Phys.*, 85:342–389, 1989.

International Series of Numerical Mathematics, Vol. 114, © 1993 Birkhäuser Verlag Basel

Regularity and Uniqueness Results for Two-Phase Miscible Flows in Porous Media

Andro Mikelić*

Abstract. We consider the Peaceman model for nonstationary two-component miscible flows through porous media. It is known that there exists a global weak solution, and we address the questions of its regularity and uniqueness. Our results hold under the assumption that the variation of the viscosity is not too large, or, speaking from a physical point of view, that the mobility ratio is not too much larger than one. More precisely, we require that the variation of the viscosity as a function of concentration multiplied by the sum of appropriate Sobolev norms of the initial and boundary data is less than a constant depending on the domain and the data. Under that condition we prove the existence of a very regular solution, the uniqueness of which is also established.

1 INTRODUCTION

We consider the regularity and uniqueness for the coupled parabolic-elliptic system

$$v = -\frac{K(x)}{\mu(c)}[\nabla p - f(x,c)] \qquad \text{in } Q_T = \Omega \times (0,T), \tag{1.1a}$$

$$\operatorname{div} v = q^I - q^P \qquad \text{in } Q_T, \tag{1.1b}$$

$$v \cdot \nu = 0 \qquad \text{on } S_T = \partial\Omega \times (0,T), \tag{1.1c}$$

$$\frac{\partial c}{\partial t} - \operatorname{div}\{D^\beta(v,c)\nabla c - vc\} + q^P c = q^I \qquad \text{in } Q_T, \tag{1.1d}$$

$$D^\beta(v,c)\nabla c \cdot \nu = g \qquad \text{on } S_T, \tag{1.1e}$$

$$c(x,0) = c_0(x) \qquad \text{in } \Omega, \tag{1.1f}$$

where Ω is a given domain in $\boldsymbol{R}^n$.

The system (1.1) models the effects of turbulent mechanical mixing of nonstationary, incompressible, two-component, miscible displacement. Usually, in petroleum engineering, this model is called the Peaceman model (see Douglas *et al.* [2], Ewing and Wheeler [3], Goyeau [5], Peaceman [11], Russell and Wheeler [12], Russell *et al.* [13] or Sammon [14] for more details).

For the Peaceman model, the dispersion tensor $D^\beta(v,c)$ is given by

$$D^\beta(v,c) = \lambda(c)I + d_\ell|v|^{1+\beta}T(v) + d_t|v|^{1+\beta}(I - T(v)), \tag{1.2a}$$

$$T_{ij}(v) = \frac{v_i v_j}{|v|^2}, \tag{1.2b}$$

*Equipe d'Analyse Numérique, Université Jean Monnet de Saint-Etienne, Faculté des Sciences et Techniques, 23, rue du Dr Paul Michelon, 42023 Saint-Etienne Cédex France and University of Zagreb, Zagreb, Croatia

where λ is a strictly positive, continuous function, d_t and d_l are positive constants, and $0 \leq \beta < 1$. The most important case is $\beta = 0$, and it is considered in a large number of papers from the point of view of numerical analysis (see Douglas *et al.* [2], Russell and Wheeler [12], Russell *et al.* [13], Sammon [14], and references therein).

In (1.1), the behavior of viscosity μ as a function of c is very important from a physical point of view because it is connected with viscous fingering. In Mikelić [9] it was shown that there exists a unique smooth solution for the stationary model, provided that μ' is not too large and data are small. Consequently, it is quite natural that, even with the knowledge that there exists at least one weak solution for (1.1) (as proved in Fabrie and Langlais [4]), we can prove its smoothness only for μ' being not too large in an appropriate norm.

It is interesting to note that in numerical simulations the most frequent form of μ is

$$\frac{1}{\mu(c)} = \frac{1}{\mu(0)}\left[(M^{\frac{1}{4}} - 1)c + 1\right]^4, \quad c \in [0,1]. \tag{1.3}$$

The number $M = \mu(0)/\mu(1)$ is the mobility ratio and physical instabilities happen for $M > 1$. In this situation, the statement "μ' is not too large" means that $M^{\frac{1}{4}} - 1$ is not too big.

In the case $M = 1$ (i.e., $\mu' = 0$) and $f(x,c) = f(x)$ the system (1.1) was considered by Sammon [14], who showed that the solution is of class C^∞. On the other hand, Fabrie and Langlais [4] demonstrated the existence of at least one weak solution $(p,v,c) \in L^\infty(0,T;H^1(\Omega)) \times L^\infty(0,T;L^2(\Omega)) \times L^\infty(Q_T) \cap L^2(0,T;H^1(\Omega))$, regardless of the magnitude of μ'. However, they did not prove any regularity and, consequently, uniqueness was left open. The natural question which we answer in this paper is under which conditions (1.1) has a smooth solution. A result developed in Fabrie and Langlais [4] implies uniqueness of smooth solutions. In the stationary case, the answer to this question is given in Mikelić [9].

We adopt the notation used in the book of Ladyženskaya *et al.* [8] for function spaces related to parabolic problems. We shall abbreviate the notation for $C^\gamma([0,T];B)$ to $C^\gamma(B)$ where B is any Banach space defined on functions over Ω.

2 STATEMENT OF THE PROBLEM AND SOME AUXILIARY RESULTS

In this section we formulate the problem (1.1) in a precise way. We assume that $\Omega \subset \boldsymbol{R}^n$, $n > 1$, is a bounded domain with a boundary $\Gamma \in C^{1,1}$. Then the problem is as follows:

Find a triple $(p,v,c) \in C^\gamma(W^{2,r}(\Omega)) \times C^\gamma(W^{1,r}(\Omega)) \times W_r^{2,1}(Q_T)$ such that

$$v = -\zeta(c)K(x)[\nabla p - f(x,c)] \quad \text{a.e. in } Q_T, \tag{2.1a}$$

$$\operatorname{div} v = q^I - q^P \quad \text{a.e. in } Q_T, \tag{2.1b}$$

$$v \cdot \nu = 0 \quad \text{a.e. on } S_T, \tag{2.1c}$$

$$\frac{\partial c}{\partial t} - \operatorname{div}\{D^\beta(v,c)\nabla c - vc\} + q^P c = q^I \quad \text{a.e. in } Q_T, \tag{2.1d}$$

$$D^\beta(v,c)\nabla c \cdot \nu = g \quad \text{a.e. on } S_T, \tag{2.1e}$$

$$c(x,0) = c_0(x) \quad \text{a.e. in } \Omega. \tag{2.1f}$$

Here D^β is given by (1.2) for $\beta \in [0,1]$. The matrix K is symmetric and such that $K \in W^{1,\infty}(\Omega)$ and $(Kx, x) \geq a|x|^2$, $\forall x \in \boldsymbol{R}^n$; ν is the outward unit normal. We assume that

$$\begin{cases} q^P, q^I \in H^{1-\frac{n+2}{r}, \frac{1}{2}-\frac{n+2}{2r}}(\overline{Q}_T), \quad q^I, q^P \geq 0 \text{ in } Q_T, \\ g \in W_\infty^{1,\frac{1}{2}}(S_T), \quad f \in W^{2,\infty}(\Omega \times \boldsymbol{R})^n, \\ \lambda \in W^{1,\infty}(\boldsymbol{R}), \quad \lambda \geq \lambda_0 > 0, \quad c_0 \in W^{2-\frac{2}{r},r}(\Omega), \quad r > 2(n+1), \end{cases} \tag{2.2}$$

together with the compatibility condition

$$\int_\Omega (q^I - q^P) = 0, \quad \forall t \in [0,T].$$

We suppose that the function $\zeta = 1/\mu$ is such that

$$\zeta \in W^{2,\infty}(\boldsymbol{R}), \quad 0 < \zeta_* \leq \zeta(y) \leq \zeta^* < +\infty, \quad \forall y \in \boldsymbol{R}, \tag{2.3}$$

when extended appropriately outside the interval $[0,1]$. Obviously, $\zeta = 1/\mu$, if based on (1.3), satisfies (2.3).

We start by recalling the properties of the dispersion tensor D^β. We have the following result from Mikelić [9].

Lemma 2.1 *For $u \in C(\overline{Q}_T)^n$ and $c \in C(\overline{Q}_T)$, the inequality*

$$(D^\beta(u,c)\xi, \xi) \leq \left[\lambda_0 + \min\{d_\ell, d_t\}|u|^{1+\beta}\right] |\xi|^2 \forall \xi \in \boldsymbol{R}^n,$$

holds. Furthermore, $D^\beta_{ij} \in W^{1,\infty}_{loc}(\boldsymbol{R} \times \boldsymbol{R})$, $\forall i,j \in \{1,\dots,n\}$.

Our next step is to establish the properties of the system (2.1*a-c*) for a given c. In order to do this, we study the auxiliary problem

$$\begin{cases} -\operatorname{div}\{\zeta(0)K(x)[\nabla\pi - G(x,t)]\} = R(x,t), \\ \zeta(0)K(x)[\nabla\pi - G]\nu = 0, \\ \displaystyle\int_\Gamma R(x,t)d\Gamma = 0, \quad \forall t \in [0,T]. \end{cases} \tag{2.4}$$

Lemma 2.2 *Let $K \in (W^{1,\infty}(\Omega))^{n^2}$ be a symmetric, positive-definite matrix, and let $G \in C^\gamma(W^{1,\infty}(\Omega))$ and $R \in C^\gamma(L^\infty(\Omega))$, $0 \leq \gamma < 1$. Then, (2.4) has a unique solution π belonging to $C^\gamma(W^{2,r}(\Omega))$ for every $r < +\infty$, such that $\int_\Omega \pi dx = 0$ a.e. for $t \in [0,T]$. Furthermore, there exists a constant $C_1 = C_1(\Omega, \|K\|_{1,\infty})$ such that*

$$\|\pi\|_{C^\gamma(W^{2,r}(\Omega))} \leq C_1 \left\{\|R\|_{C^\gamma(L^r(\Omega))} + \|G\|_{C^\gamma(W^{1,r}(\Omega))}\right\}. \tag{2.5}$$

Proof: See, e.g., Grisvard [6] or Nečas [10]. Note that time is just a parameter. ■

Lemma 2.3 *Let ζ satisfy (2.3), and let the functions s_1 and s_2 belong to $W^{1,\infty}(\Omega)$. Then,*

$$\|\zeta(s_1) - \zeta(s_2)\|_{W^{1,\infty}(\Omega)} \leq (n+1)\left[1 + \|s_2\|_{W^{1,\infty}(\Omega)}\right] \|\zeta\|_{W^{1,\infty}(\boldsymbol{R})} \|s_1 - s_2\|_{W^{1,\infty}(\Omega)}. \tag{2.6}$$

Let $C_{tr}(\Omega, r)$ be the norm of the trace map:

$$\|\phi\|_{W^{1-\frac{1}{r},r}(\Gamma)} \leq C_{tr}(\Omega, r)\|\phi\|_{W^{1,r}(\Omega)}, \quad \forall \phi \in W^{1,r}(\Omega). \tag{2.7}$$

Set

$$C_2 = \max(1, C_1)\max\{1, (1 + C_{tr}(\Omega; r))\|K\|_{W^{1,\infty}(\Omega)}(n+1)^2\}.$$

For $h \in C(\overline{Q}_T)$, $\varphi \in W^{1,\infty}(\Omega \times \boldsymbol{R})^n$, and $u \in C(W^{1,\infty}(\Omega))$, let

$$\begin{aligned} A_1(h, \varphi, u) &= \|h\|_{C(L^\infty(\Omega))} + \|\varphi\|_{W^{1,\infty}(\Omega; L^\infty(R))^n} \\ &\quad + \left\|\frac{\partial \varphi}{\partial s}\right\|_{L^\infty(\Omega \times R)^n} \|u\|_{C(W^{1,\infty}(\Omega))}. \end{aligned} \tag{2.8}$$

Furthermore, set

$$A_2(h, \varphi, u) = \max\left\{1, \|\zeta\|_{L^\infty(R)}\right\} \frac{5}{2} C_2 |\Omega|^{\frac{1}{r}} A_1(h, \varphi, u)$$

and

$$\psi(y) = (1+y)/(1-y).$$

Lemma 2.4 *Let the assumptions of Lemma 2.2 hold, and let $s \in C^\gamma(W^{1,\infty}(\Omega))$. If $p \in C^\gamma(W^{2,r}(\Omega))$, normalized so that $\int_\Omega p = 0$, is a solution of*

$$\left\{ \begin{array}{ll} -div\{\zeta(s)K(x)[\nabla p - f(x,s)\} = q^I - q^P & in\ \Omega, \\ \zeta(s)K(x)[\nabla p - f(x,s)] \cdot \nu = 0 & on\ \Gamma, \end{array} \right. \tag{2.9}$$

for $t \in [0, T]$ and if

$$C_2\|\zeta'\|_{L^\infty(R)}\|s\|_{C(W^{1,\infty}(\Omega))} < 1, \tag{2.10}$$

then

$$\|p\|_{C(W^{2,r}(\Omega))} \leq C_2 A_1(q^I - q^P, f, s)\psi\left(C_2\|\zeta'\|_{L^\infty(R)}\|s\|_{C(W^{1,\infty}(\Omega))}\right)|\Omega|^{\frac{1}{r}} \tag{2.11}$$

and

$$\begin{aligned} \|p\|_{C^\gamma(W^{2,r}(\Omega))} &\leq |\Omega|^{\frac{1}{r}} C_2 A_1(q^I - q^P, f, s) \cdot \\ &\quad \cdot \left\{\psi(C_2\|\zeta'\|_{L^\infty}\|s\|_{C(W^{1,\infty})}) + 2C_2 \frac{\|\zeta'\|_{L^\infty}\|s\|_{C^\gamma(W^{1,\infty})}}{[1 - C_2\|\zeta'\|_{L^\infty}\|s\|_{C(W^{1,\infty})}]^2}\right\} \\ &+ |\Omega|^{\frac{1}{r}} C_2 \psi(C_2\|\zeta'\|_{L^\infty}\|s\|_{C(W^{1,\infty})}) \cdot \\ &\quad \cdot \left[A_1\left(\frac{\widehat{q}^I - \widehat{q}^P}{\tau^\gamma}, f, \frac{\widehat{s}}{\tau^\gamma}\right) + \left\|\frac{\partial^2 f}{\partial s^2}\right\|_{L^\infty(\Omega \times R)} \|s\|_{C^\gamma(W^{1,\infty})}\right], \end{aligned} \tag{2.12}$$

where $\widehat{u} = u(t+\tau) - u(t)$, $\tau > 0$.

Proof: We rewrite (2.9) in the form

$$\begin{cases} -\operatorname{div}\{\zeta(0)K(x)[\nabla p - f(x,s)]\} \\ \qquad = q^I - q^P - \operatorname{div}\{[\zeta(0) - \zeta(s)]K[\nabla p - f(x,s)]\} & \text{in } \Omega, \\ \zeta(0)K(x)[\nabla p - f]\cdot\nu = [\zeta(0) - \zeta(s)]K[\nabla p - f]\cdot\nu & \text{on } \Gamma, \end{cases} \tag{2.13}$$

and introduce the auxiliary function $\widetilde{v} = -K(x)(\nabla p - f)$. It is important to note that

$$\|\widetilde{v}\|_{C(W^{1,r}(\Omega))} \leq (n+1)\|K\|_{W^{1,\infty}(\Omega)}\Big[\|p\|_{C(W^{2,r}(\Omega))} + \|f\|_{C(W^{1,r}(\Omega))}\Big].$$

A direct consequence of Lemma 2.2 and the inequalities (2.6), (2.7), and (2.13) is that

$$\begin{aligned} \|p\|_{C(W^{2,r}(\Omega))} \leq C_1\Big\{&\|q^I - q^P\|_{C(L^r(\Omega))} + \|f\|_{C(W^{1,r}(\Omega))} \\ &+ (n+1)^2\|s\|_{C(W^{1,\infty}(\Omega))}\|\zeta'\|_{L^\infty(R))}\|K\|_{W^{1,\infty}(\Omega)}\cdot \\ &\cdot\Big[\|p\|_{C(W^{2,r}(\Omega))} + \|f\|_{C(W^{1,r}(\Omega))}\Big][1 + C_{tr}(\Omega,r)]\Big\}. \end{aligned} \tag{2.14}$$

Now, (2.14) can be written in the form

$$\begin{aligned} &\Big\{1 - C_2\|\zeta'\|_{L^\infty(R)}\|s\|_{C(W^{1,\infty}(\Omega))}\Big\}\|p\|_{C(W^{2,r}(\Omega))} \\ &\quad\leq C_2\Big\{\|q^I - q^P\|_{C(L^r(\Omega))} + \Big[1 + \|\zeta'\|_{L^\infty(R)}\|s\|_{C(W^{1,\infty}(\Omega))}\Big]\|f\|_{C(W^{1,r}(\Omega))}\Big\}. \end{aligned}$$

This inequality implies that

$$\begin{aligned} \|p\|_{C(W^{2,r}(\Omega))} \leq C_2&\frac{1 + \|\zeta'\|_{L^\infty(R)}\|s\|_{C(W^{1,\infty}(\Omega))}C_2}{1 - \|\zeta'\|_{L^\infty(R)}\|s\|_{C(W^{1,\infty}(\Omega))}C_2}\cdot \\ &\cdot\Big\{\|q^I - q^P\|_{C(L^r(\Omega))} + \|f\|_{C(W^{1,r}(\Omega)}\Big\}. \end{aligned} \tag{2.15}$$

Now, we estimate $f(x,s)$:

$$\begin{aligned} \|f\|_{C(W^{1,r}(\Omega))} &\leq \|f\|_{W^{1,r}(\Omega,L^\infty(R))} + \left\|\frac{\partial f}{\partial s}\right\|_{L^\infty(\Omega\times R)}\|\nabla s\|_{C(L^r(\Omega))} \\ &\leq |\Omega|^{\frac{1}{r}}[A_1(q^I - q^P, f, s) - \|q^I - q^P\|_{C(L^\infty(\Omega))}]. \end{aligned} \tag{2.16}$$

Substituting (2.16) into (2.15) gives (2.11).

Our next step is to derive (2.12). We start with the "linearized" equation analogous to (2.13) for $\widehat{p} = p(t+\tau) - p(t)$:

$$\begin{aligned} -\operatorname{div}\{\zeta(0)K(x)[\nabla\widehat{p} - \widehat{f})\} = \widehat{q}^I - \widehat{q}^P - \operatorname{div}\Big\{&[\zeta(s(t+\tau)) \\ - \zeta(s(t))]K[\nabla p(t+\tau) - f(t+\tau)]\Big\} &+ \operatorname{div}\{[\zeta(0) - \zeta(s(t)]K[\nabla\widehat{p} - \widehat{f}]\}. \end{aligned}$$

Therefore,

$$\left\| \frac{p(t+\tau)-p(t)}{\tau^\gamma} \right\|_{C(W^{2,r}(\Omega))} \leq C_2 \psi(C_2 \|\zeta'\|_{L^\infty(R)} \|s\|_{C(W^{1,\infty}(\Omega))}) \cdot$$
$$\cdot \Big[A_1\Big(\frac{\widehat{q}^I - \widehat{q}^P}{\tau^\gamma}, f, \frac{\widehat{s}}{\tau^\gamma}\Big) + \left\| \frac{\partial^2 f}{\partial s^2} \right\|_{L^\infty(\Omega\times R)} \left\| \frac{\widehat{s}}{\tau^\gamma} \right\|_{C(W^{1,r}(\Omega)}\Big] |\Omega|^{\frac{1}{r}}$$
$$+ \frac{1}{1 - C_2 \|\zeta'\|_{L^\infty(R)} \|s\|_{C(W^{1,\infty}(\Omega))}} \cdot \tag{2.17}$$
$$\cdot \left\| \Big[\frac{\zeta(s(t+\tau)) - \zeta(s(t))}{\tau^\gamma} \Big] K[\nabla p(t+\tau) - f(t+\tau)] \right\|_{C(W^{1,r}(\Omega))}.$$

Let us estimate the last term:

$$\left\| \Big[\frac{\zeta(s(t+s)) - \zeta(s(t))}{\tau^\gamma} \Big] K[\nabla p(t+\tau) - f(t+\tau)] \right\|_{C(W^{1,r}(\Omega))} \tag{2.18}$$
$$\leq \|\zeta'\|_{L^\infty(R)} \|K\|_{W^{1,\infty}(\Omega)} n^2(n+1) \cdot$$
$$\cdot \Big[\|p\|_{C(W^{2,r}(\Omega))} + \|f\|_{C(W^{1,r}(\Omega))} \Big] \cdot \left\| \frac{\widehat{s}}{\tau^\gamma} \right\|_{C(W^{1,\infty}(\Omega))}$$
$$\leq C_2 \|\zeta'\|_{L^\infty(R)} \left\| \frac{\widehat{s}}{\tau^\gamma} \right\|_{C(W^{1,\infty}(\Omega))} \cdot$$
$$\cdot \Big[C_2 A_1(q^I - q^P, f, s) \psi(C_2 \|\zeta'\|_{L^\infty(R)} \|s\|_{C(W^{1,\infty}(\Omega))}) |\Omega|^{\frac{1}{r}}$$
$$+ |\Omega|^{\frac{1}{r}} (A_1(q^I - q^P, f, s) - \|q^I - q^P\|_{C(L^\infty(\Omega)}\Big].$$

Finally, substituting (2.18) into (2.17) gives

$$\|p\|_{C^\gamma(W^{2,r}(\Omega))} \leq C_2 \Big\{ A_1(q^I - q^P, f, s) |\Omega|^{\frac{1}{r}} \psi(C_2 \|\zeta'\|_{L^\infty(R)} \|s\|_{C(W^{1,\infty})})$$
$$+ \psi(C_2 \|\zeta'\|_{L^\infty} \|s\|_{C(W^{1,\infty})}) \cdot$$
$$\cdot \Big[A_1\Big(\frac{\widehat{q}^I - \widehat{q}^P}{\tau^\gamma}, f, \frac{\widehat{s}}{\tau^\gamma}\Big) + \left\| \frac{\partial^2 f}{\partial s^2} \right\|_{L^\infty(\Omega\times R)} \left\| \frac{\widehat{s}}{\tau^\gamma} \right\|_{C(W^{1,\infty})}\Big] |\Omega|^{\frac{1}{r}}$$
$$+ \frac{2C_2 \|\zeta'\|_{L^\infty} \left\| \frac{\widehat{s}}{\tau^\gamma} \right\|_{C(W^{1,\infty})} |\Omega|^{\frac{1}{r}}}{[1 - C_2 \|\zeta'\|_{L^\infty(R)} \|s\|_{C(W^{1,\infty})}]^2} \cdot A_1(q^I - q^P, f, s) \Big\},$$

which implies (2.12). ∎

Lemma 2.5 *Let the assumptions from Lemma 2.4 hold, and let v be given by*

$$v = -\zeta(s) K(x)(\nabla p - f).$$

Then,

$$\|v\|_{C(W^{1,r}(\Omega))} \leq A_2(q^I - q^P, f, s) \psi(C_2 \|\zeta'\|_{L^\infty} \|s\|_{C(W^{1,\infty})}),$$
$$\|v\|_{C^\gamma(W^{1,r}(\Omega))} \leq A_2(q^I - q^P, f, s) \psi(C_2 \|\zeta'\|_{L^\infty} \|s\|_{C(W^{1,\infty})}) \cdot \tag{2.19}$$
$$\cdot [1 + \|\zeta'\|_{L^\infty} \|s\|_{C^\gamma(W^{1,\infty})} + \psi(C_2 \|s\|_{C^\gamma(W^{1,\infty})} \|\zeta'\|_{L^\infty})].$$

Remark. Let us simplify estimate (2.19). We have

$$\|v\|_{C^\gamma(W^{1,r})} \leq \frac{5}{2} C_2 |\Omega|^{\frac{1}{r}} \max\{1, \|\zeta\|_{L^\infty}\} \Big[\|q^I - q^P\|_{C^\gamma(L^\infty)} + \|f\|_{W^{1,\infty}(L^\infty)} + \Big(\Big\| \frac{\partial^2 f}{\partial s^2} \Big\|_{L^\infty(\Omega\times R)} + \Big\| \frac{\partial f}{\partial s} \Big\|_{L^\infty(\Omega\times R)} \Big) \|s\|_{C^\gamma(W^{1,\infty})} \Big] \frac{3}{[1 - C_2 \|\zeta'\|_{L^\infty} \|s\|_{C^\gamma(W^{1,\infty})}]}.$$

Now, we introduce

$$A_3(h, \varphi) = \|h\|_{C^\gamma(L^\infty)} + \|\varphi\|_{W^{1,\infty}(\Omega; L^\infty(R))}, \ \forall h \in C^\gamma(L^\infty), \ \forall \varphi \in W^{1,\infty}(\Omega, L^\infty(\boldsymbol{R})),$$

$$C_3 = \frac{15}{3} C_2 |\Omega|^{\frac{1}{r}} \max\{1, \|\zeta\|_{L^\infty}\},$$

$$A_4(f) = \frac{1}{C_2} \Big[\Big\| \frac{\partial^2 f}{\partial s^2} \Big\|_{L^\infty(\Omega\times R)} + \Big\| \frac{\partial f}{\partial s} \Big\|_{L^\infty(\Omega\times R)} \Big].$$

Then, finally

$$\|v\|_{C^\gamma(W^{1,r})} \leq C_3 \frac{A_3(q^I - q^P, f) + A_4(f) C_2 \|s\|_{C^\gamma(W^{1,\infty})}}{1 - C_2 \|\zeta'\|_{L^\infty} \|s\|_{C^\gamma(W^{1,\infty})}}.$$

If the function $\widetilde{\psi}(y)$ is defined by

$$\widetilde{\psi}(y) = C_3 \frac{A_3(q^I - q^P, f) + A_4(f) y / \|\zeta'\|_{L^\infty}}{(1-y)^2},$$

then

$$\|v\|_{C^\gamma(W^{1,r})} \leq \widetilde{\psi}(y) = \widetilde{\psi}(C_2 \|\zeta'\|_{L^\infty} \|s\|_{C^\gamma(W^{1,r})}). \tag{2.20}$$

Proof: We use the inequality (2.13). We have

$$\begin{aligned}
\|v\|_{C(W^{1,r}(\Omega))} \leq\ & \|\zeta\|_{L^\infty(R)} (n+1)^2 \|K\|_{W^{1,\infty}(\Omega)^{n^2}} \Big[\|p\|_{C(W^{2,r}(\Omega))} + \|f\|_{C(W^{1,r}(\Omega))} \Big] \\
& + \|\zeta'\|_{L^\infty(R)} \|s\|_{C(W^{1,\infty}(\Omega))} (n+1) \|K\|_{L^\infty(\Omega)} \cdot \\
& \cdot \Big[\|p\|_{C(W^{1,r}(\Omega))} + \|f\|_{C(L^r(\Omega))} \Big] \\
\leq\ & C_2 \Big[\|\zeta\|_{L^\infty(R)} + \|\zeta'\|_{L^\infty} \|s\|_{C(W^{1,\infty})} C_2 \Big] \|p\|_{C(W^{2,r})} \\
& + C_2 \|\zeta\|_{L^\infty} \|f\|_{C(W^{1,r})} + C_2 \|\zeta'\|_{L^\infty} \|s\|_{C(W^{1,\infty})} \|f\|_{C(L^r)} \\
\leq\ & C_2 \Big[\|\zeta\|_{L^\infty(R)} + \|\zeta'\|_{L^\infty} \|s\|_{C(W^{1,\infty})} C_2 \Big] \cdot \\
& \cdot C_2 |\Omega|^{\frac{1}{r}} A_1(q^I - q^P, f, s) \psi(C_2 \|\zeta'\|_{L^\infty} \|s\|_{C(W^{1,\infty})}) \\
& + C_2 \|\zeta'\|_{L^\infty} \|s\|_{C(W^{1,\infty})} \|f\|_{L^\infty} |\Omega|^{\frac{1}{r}} \\
& + C_2 \|\zeta\|_{L^\infty} \Big[\|f\|_{W^{1,\infty}(\Omega; L^\infty(R))} + \Big\| \frac{\partial f}{\partial s} \Big\|_{L^\infty(\Omega\times R)} \|s\|_{C(W^{1,\infty})} \Big] |\Omega|^{\frac{1}{r}} \\
\leq\ & C_2^2 \mathrm{Max}\, \{1, \|\zeta\|_{L^\infty(R)}\} |\Omega|^{\frac{1}{r}} A_1(q^I - q^P, f, s) \frac{4}{1 - C_2 \|\zeta'\|_{L^\infty} \|s\|_{C(W^{1,\infty})}}
\end{aligned}$$

$$\begin{aligned}
&+\|f\|_{L^\infty(\Omega\times R)}|\Omega|^{\frac{1}{r}}+C_2\|\zeta\|_{L^\infty}\|f\|_{W^{1,\infty}(\Omega;L^\infty(R))}|\Omega|^{\frac{1}{r}}\\
&+C_2\|\zeta\|_{L^\infty}|\Omega|^{\frac{1}{r}}\left\|\frac{\partial f}{\partial s}\right\|_{L^\infty(\Omega\times R)}\|s\|_{C(W^{1,\infty})}\\
\leq\ &\frac{5}{2}C_2\max\{1,\|\zeta\|_{L^\infty(R)}\}|\Omega|^{\frac{1}{r}}A_1(q^I-q^P,f,s)\psi(C_2\|\zeta\|_{L^\infty}\|s\|_{C(W^{1,\infty})}),
\end{aligned}$$

which implies (2.19).

Now, let us derive an estimate for the Hölder norm in time. Let

$$\widehat{v}=-\zeta(s(t+\tau))K(\nabla\widehat{p}-\widehat{f})+(\zeta(s(t))-\zeta(s(t+\tau)))K(x)(\nabla p(t)-f(t));$$

then,

$$\begin{aligned}
\left\|\frac{\widehat{v}}{\tau^\gamma}\right\|_{C(W^{1,r})}\leq\ &\|\zeta'\|_{L^\infty}\left\|\frac{\widehat{s}}{\tau^\gamma}\right\|_{C(W^{1,\infty})}C_2|\Omega|^{\frac{1}{r}}A_1(q^I-q^P,f,s)\psi(C_2\|\zeta'\|_{L^\infty}\|s\|_{C(W^{1,r})})\\
&+C_2\Big[\|\zeta\|_{L^\infty}+\|\zeta'\|_{L^\infty}\|s\|_{C(W^{1,\infty})}C_2\Big]\left[\left\|\frac{\widehat{p}}{\tau^\gamma}\right\|_{C(W^{2,r})}+\left\|\frac{\widehat{f}}{\tau^\gamma}\right\|_{C(W^{1,r})}\right]\\
\leq\ &\|\zeta'\|_{L^\infty}\left\|\frac{\widehat{s}}{\tau^\gamma}\right\|_{C(W^{1,r})}C_2|\Omega|^{\frac{1}{r}}A_1(q^I-q^P,f,s)\psi(C_2\|\zeta'\|_{L^\infty}\|s\|_{C(W^{1,\infty})})\\
&+2C_2\max\{1,\|\zeta\|_{L^\infty}\}\Big\{|\Omega|^{\frac{1}{r}}C_2A_1(q^I-q^P,f,s)\cdot\\
&\cdot\Big[\psi(C_2\|\zeta'\|_{L^\infty}\|s\|_{C(W^{1,\infty})})+\frac{1}{2}C_2\|\zeta'\|_{L^\infty}\|s\|_{C(W^{1,\infty})}\psi^2\Big]\\
&+|\Omega|^{\frac{1}{r}}C_2\psi\Big[A_1\Big(\frac{\widehat{q}^I-\widehat{q}^P}{\tau^\gamma}f,s,\frac{\widehat{s}}{\tau^\gamma}\Big)+\left\|\frac{\partial^2 f}{\partial s^2}\right\|_{L^\infty(\Omega\times R)}\|s\|_{C^\gamma(W^{1,\infty})}\Big]\\
&+\left\|\frac{\partial f}{\partial s}\right\|_{L^\infty}\left\|\frac{\widehat{s}}{\tau^\gamma}\right\|_{L^\infty}|\Omega|^{\frac{1}{r}}+\left\|\frac{\partial^2 f}{\partial s^2}\right\|_{L^\infty}\left\|\frac{\nabla\widehat{s}}{\tau^\gamma}\right\|_{L^\infty}|\Omega|^{\frac{1}{r}}\Big\}\\
\leq\ &2C_2|\Omega|^{\frac{1}{r}}\max\{1,\|\zeta\|_{L^\infty}\}A_1(q^I-q^P,f,s)\psi(C_2\|\zeta'\|_{L^\infty}\|s\|_{C(W^{1,\infty})})\cdot\\
&\cdot\Big[\|\zeta'\|_{L^\infty}\left\|\frac{\widehat{s}}{\tau^\gamma}\right\|_{C(W^{1,\infty})}+1\Big]+A_1(q^I-q^P,f,s)\psi^2(C_2\|\zeta'\|_{L^\infty}\|s\|_{C(W^{1,\infty})})\\
&+2|\Omega|^{\frac{1}{r}}C_2\psi\Big(C_2\|\zeta'\|_{L^\infty}\|s\|_{C(W^{1,\infty})}\Big)\Big\{A_1\Big(\frac{\widehat{q}^I-\widehat{q}^P}{\tau^\gamma},f,\frac{\widehat{s}}{\tau^\gamma}\Big)\\
&+\left\|\frac{\partial^2 f}{\partial s^2}\right\|_{L^\infty(\Omega\times R)}\|s\|_{C^\gamma(W^{1,\infty})}\Big\},
\end{aligned}$$

which implies (2.20). ■

Set

$$C_E=\|K\|_{L^\infty}\|\zeta\|_{L^\infty}(C_B+1)+\frac{1}{\min\zeta}(A_3(q^I-q^P,f)+1)$$

and

$$C_D=C_A\|q^I-q^P\|_{W^{\sigma,r}(L^r)}+|\Omega|^{\frac{1}{r}}\|f\|_{L^\infty}. \tag{2.21}$$

Lemma 2.6 *Let the assumptions of Lemma 2.5 hold, and let*

$$s\in C^\gamma(W^{1,\infty})\cap W^{\sigma,r}(0,T;L^r(\Omega))$$

for $0 < \sigma < 1$. *Then,* $v \in W^{\sigma,r}(0,T;L^r(\Omega)) \cap C^\gamma(W^{1,r})$, *and*

$$\|v\|_{W^{\sigma,r}(L^r)} \leq C_E \|s\|_{W^{\sigma,r}(L^r)} \left(\left\| \frac{\partial f}{\partial s} \right\|_{L^\infty} + \|\zeta'\|_{L^\infty} \right) \psi(C_2 \|\zeta'\|_{L^\infty} \|s\|_{C(W^{1,\infty})}) + C_D. \quad (2.22)$$

Proof: In the smooth case, $\left\{ \frac{\partial p}{\partial t}, \frac{\partial v}{\partial t} \right\}$ is a unique solution of

$$\begin{cases} \frac{\partial v}{\partial t} = -K\zeta(s)\nabla \frac{\partial p}{\partial t} - \frac{\partial s}{\partial t} K \left[\zeta'(\nabla p - f) - \zeta \frac{\partial f}{\partial s} \right] & \text{in } \Omega, \\ \operatorname{div} \frac{\partial v}{\partial t} = \frac{\partial}{\partial t}(q^I - q^P) & \text{in } \Omega, \\ \frac{\partial v}{\partial t} \cdot \nu = 0 & \text{on } \Gamma. \end{cases}$$

Then,

$$\left\| \frac{\partial p}{\partial t} \right\|_{L^r(W^{1,r})} \leq C_A \left\{ \left\| \frac{\partial s}{\partial t} K \left[\zeta'(\nabla p - f) - \zeta \frac{\partial f}{\partial s} \right] \right\|_{L^r(L^r)} + \left\| \frac{\partial}{\partial t}(q^I - q^P) \right\|_{L^r(L^r)} \right\},$$

where C_A depends only on Ω, n, r, $\|K\|_{C(\Omega)}$, $\|\zeta\|_{C(R)}$ and $\|s\|_{C(\overline{Q})}$ (see Antontsev *et al.* [1], pp. 220–226). Furthermore,

$$\|p\|_{L^r(W^{1,r})} \leq C_A \left\{ \|K\zeta f\|_{L^r(L^r)} + \|q^I - q^P\|_{L^r(L^r)} \right\},$$

and by interpolation we get

$$\begin{aligned} \|p\|_{W^{\sigma,r}(W^{1,r})} \leq \; & C_A \Big\{ \|K\zeta f\|_{W^{\sigma,r}(L^r)} \quad (2.23) \\ & + \|K(\nabla p - f)\|_{C(L^\infty)} \|\zeta\|_{W^{\sigma,r}(L^r)} + \|q^I - q^P\|_{W^{\sigma,r}(L^r)} \Big\} \end{aligned}$$

(see also Triebel [16], for an argument with finite differences). Now, (2.23) implies that

$$\begin{aligned} \|p\|_{W^{\sigma,r}(W^{1,r})} \leq \; & C_A \Big\{ \|K\|_{L^\infty} \|\zeta\|_{L^\infty} \left\| \frac{\partial f}{\partial s} \right\|_{L^\infty} \|s\|_{W^{\sigma,r}(L^r)} \\ & + \widetilde{C}_{emb} \frac{1}{\min \zeta} \|\zeta'\|_{L^\infty} \|s\|_{W^{\sigma,r}(L^r)} \|v\|_{C(W^{1,r})} + \|q^I - q^P\|_{W^{\sigma,r}(L^r)} \Big\} \\ \leq \; & C_A \Big\{ \Big[\|K\|_{L^\infty} \|\zeta\|_{L^\infty} \left\| \frac{\partial f}{\partial s} \right\|_{L^\infty} + \frac{\widetilde{C}_{emb}}{\min \zeta} \|\zeta'\|_{L^\infty} A_2(q^I - q^P, f, s) \cdot \\ & \cdot \psi(C_2 \|\zeta'\|_{L^\infty} \|s\|_{C(W^{1,\infty})}) \Big] \|s\|_{W^{\sigma,r}(L^r)} + \|q^I - q^P\|_{W^{\sigma,r}(L^r)} \Big\} \\ \leq \; & C_A \Big\{ \Big[\|K\|_{L^\infty} \|\zeta\|_{L^\infty} \left\| \frac{\partial f}{\partial s} \right\|_{L^\infty} \\ & + \frac{\widetilde{C}_{emb}}{\min \zeta} \|\zeta'\|_{L^\infty} \left(A_3(q^I - q^P, f) + \left\| \frac{\partial f}{\partial s} \right\|_{L^\infty} \|s\|_{C(W^{1,\infty})} \right) \cdot \\ & \cdot \psi(C_2 \|\zeta'\|_{L^\infty} \|s\|_{C(W^{1,\infty})}) \Big] \|s\|_{C(WL^{1,\infty})} + \|q^I - q^P\|_{W^{\sigma,r}(L^r)} \Big\}. \end{aligned}$$

Set

$$C_B = 2C_A \max\left\{ \|K\|_{L^\infty}\|\zeta\|_{L^\infty}, \frac{\tilde{C}_{emb}}{\min\zeta} A_3(q^I - q^P, f), \frac{\tilde{C}_{emb}}{C_2 \min\zeta} \right\}. \tag{2.24}$$

Then,

$$\begin{aligned} \|p\|_{W^{\sigma,r}(W^{1,r})} &\leq C_B \|s\|_{W^{\sigma,r}(L^r)} \left(\left\| \frac{\partial f}{\partial s} \right\|_{L^\infty} + \|\zeta'\|_{L^\infty} \right) \psi(C_2 \|\zeta'\|_{L^\infty} \|s\|_{C(W^{1,\infty})}) \\ &\quad + C_A \|q^I - q^P\|_{W^{\sigma,r}(L^r)}. \end{aligned}$$

Using (2.24) and the estimate for the pressure p gives us an estimate for the velocity:

$$\begin{aligned} \|v\|_{W^{\sigma,r}(L^r)} &= \|K\zeta(\nabla p - f)\|_{W^{\sigma,r}(L^r)} \\ &\leq \|K\|_{L^\infty}\|\zeta\|_{L^\infty}(\|\nabla p\|_{W^{\sigma,r}(L^r)} + \|f\|_{W^{\sigma,r}(L^r)}) \\ &\quad + \|v\|_{C(L^\infty)} \frac{1}{\min\zeta} \|\zeta'\|_{L^\infty} \|s\|_{W^{\sigma,r}(L^r)} \\ &\leq \|K\|_{L^\infty}\|\zeta\|_{L^\infty} \Big\{ C_B \ \|s\|_{W^{\sigma,r}(L^r)} \psi(C_2 \|\zeta'\|_{L^\infty} \|s\|_{C(W^{1,\infty})}) \cdot \\ &\quad \cdot \left(\left\| \frac{\partial f}{\partial s} \right\|_{L^\infty} + \|\zeta'\|_{L^\infty} \right) + C_A \|q^I - q^P\|_{W^{\sigma,r}(L^r)} + \|f\|_{L^\infty} |\Omega|^{\frac{1}{r}} \\ &\quad + \left\| \frac{\partial f}{\partial s} \right\|_{L^\infty} \|s\|_{W^{\sigma,r}(L^r)} \Big\} + \|\zeta'\|_{L^\infty} \|s\|_{W^{\sigma,r}(L^r)} \frac{1}{\min\zeta} \Big(A_3(q^I - q^P, f) \\ &\quad + \left\| \frac{\partial f}{\partial s} \right\|_{L^\infty} \|s\|_{C(W^{1,\infty})} \Big) \psi(C_2 \|s\|_{W^{\sigma,r}(L^r)} \|\zeta'\|_{L^\infty}) \\ &\leq \|s\|_{W^{\sigma,r}(L^r)} \left(\left\| \frac{\partial f}{\partial s} \right\|_{L^\infty} + \|\zeta'\|_{L^\infty} \right) \psi(C_2 \|\zeta'\|_{L^\infty} \|s\|_{C(W^{1,\infty})}) \cdot \\ &\quad \cdot \left\{ \|K\|_{L^\infty}\|\zeta\|_{L^\infty}(C_B + 1) + \frac{1}{\min\zeta}(A_3(q^I - q^P, f) + 1) \right\} \\ &\quad + C_A \|q^I - q^P\|_{W^{\sigma,r}(L^r)} + |\Omega|^{\frac{1}{r}} \|f\|_{L^\infty}. \end{aligned}$$

This completes the proof of the lemma. ■

After studying the equation for pressure with a fixed concentration we turn to the problem (2.1*d-f*), with a given velocity and given diffusion term λ. Hence, we study the problem

$$\begin{cases} \dfrac{\partial c}{\partial t} - \operatorname{div}\{D^\beta(u,s)\nabla c - uc\} + q^P c = q^I & \text{in } Q_T, \\ D^\beta(u,s)\nabla c \cdot \nu = g & \text{on } S_T, \\ c(x,0) = c_0 & \text{in } \Omega, \end{cases} \tag{2.25}$$

where $u \in C^\gamma(W^{1,r})$, and $s \in C^\gamma(W^{1,r}) \cap W^{\sigma,r}(L^r)$ are specified functions with $u \cdot \nu = 0$ on S_T. However, to get a smooth solution for (2.25) we need very precise compatibility conditions. Consequently, we suppose that $s(x,0) = c_0$ in Ω. Furthermore, let the pair $\{u_0, p_0\} \in W^{1,r}(\Omega) \times W^{2,r}(\Omega)$ be the solution of the problem

$$\begin{cases} u_0 = \zeta(c_0)K(x)[f(x,c_0) - \nabla p_0] & \text{in } \Omega, \\ \operatorname{div} u_0 = q^I(x,0) - q^P(x,0) & \text{in } \Omega, \\ u_0 \cdot \nu = 0 & \text{on } \Gamma. \end{cases} \tag{2.26}$$

Then, the compatibility condition is

$$D^{\beta}(u_0, c_0)\nabla c_0 \cdot \nu = g(x,0) \in W^{1-\frac{3}{r},r}(\Omega), \qquad \forall r < +\infty. \tag{2.27}$$

Lemma 2.7 *Let*

$$s \in C^{\gamma}(W^{1,\infty}(\Omega)) \cap W^{\sigma,r}(0,T;L^r(\Omega)),$$

where

$$\frac{1}{2} + \frac{n+2}{2r} < \sigma < 1, \quad 0 < \gamma < 1, \quad r > 2(n+1),$$

and let $s(x,0) = c_0 \in W^{2-\frac{2}{r},r}(\Omega)$. *Further, let*

$$\begin{aligned} u &\in C^{\gamma}(W^{1,r}(\Omega)), \\ div\, u &= q^I - q^P \text{ in } Q_T, \\ u \cdot \nu &= 0 \text{ on } S_T. \end{aligned}$$

Finally, let the compatibility condition (2.27) and the smoothness assumption (2.2) hold.

Then, (2.25) has a unique solution $c \in W_r^{2,1}(Q_T)$, *for any* $r < +\infty$, *and the following inequality holds:*

$$\begin{aligned} \|c\|_{W_r^{2,1}(Q_T)} \leq\; & \Phi\Big(\|u\|_{C^{\gamma}(W^{1,r})} + \Big\|\frac{\partial \lambda}{\partial s}\Big\|_{L^\infty} \|s\|_{C^{\gamma}(W^{1,r})}\Big) \cdot \\ & \cdot \Big\{ \|c_0\|_{W^{2-\frac{2}{r},r}(\Omega)} + \|q^I\|_{L^r(Q_T)} + \|g\|_{W_r^{1-\frac{1}{r},\frac{1}{2}-\frac{1}{2r}}(S_T)} \Big\}, \end{aligned} \tag{2.28}$$

where the function Φ *is monotone increasing in its argument and also depends on* n, Ω, $\|D^{\beta}\|_{W^{1,\infty}}$, $\|q^I\|_{L^r(Q_T)}$, $\|q^P\|_{L^r(Q_T)}$, $\|c_0\|_{W^{2-\frac{2}{r},r}(\Omega)}$, *and* $\|u_0\|_{W^{1,r}(\Omega)}$.

Proof: Results in Ladyženskaya *et al.* [8] imply existence and uniqueness of a unique weak solution in $W_2^{1,\frac{1}{2}}(Q_T)$. We will use parabolic regularity to conclude that $c \in W_r^{2,1}(Q_T)$; an appropriate general parabolic regularity theory for parabolic systems is developed in Solonnikov [15]. The coefficients of the differential operator are sufficiently smooth. We need to check the smoothness of the coefficients of the boundary operator. Theorem 1.2 in Solonnikov [15], pp. 161-162, gives smoothness conditions for the coefficients of the Neumann boundary operator, and, for $r > n+2$, those conditions are fulfilled if $D_{ij}^{\beta} \in W_r^{1-\frac{1}{r},\frac{1}{2}-\frac{1}{2r}}(\Gamma)$.

We now prove that D_{ij}^{β} has such regularity. We have $v \in W^{\sigma,r}(L^r) \cap W^{\overline{\lambda},r}(W^{1,r})$ for $\overline{\lambda} < \dfrac{1}{2} - \dfrac{n+2}{r}$ and $\sigma < 1$. A simple interpolation argument shows that $v \in W^{(1-\theta)\sigma+\theta\overline{\lambda},r}(W^{\theta,r})$ for $0 \leq \theta \leq 1$. We choose $\theta = \dfrac{\sigma}{1+\sigma-\overline{\lambda}}$ and get $v \in W^{\alpha,r}(Q_T)$ with $\alpha = \dfrac{\sigma}{1+\sigma-\overline{\lambda}}$. Let $1 > \sigma > \dfrac{1}{2} + \dfrac{n+2}{2r}$. Then, $\alpha > \dfrac{1}{2}$ and $\alpha > \dfrac{n+1}{r}$ for $r > 2(n+1)$. Now, Theorem 3, p. 449, in Hanouzet [7] implies that $D_{ij}^{\beta} \in W^{\frac{1}{2}-\frac{1}{2r}}(L^r)$; $v \in C^{\gamma}(W^{1-\frac{1}{r},r}(\Gamma))$, and $s \in C^{\gamma}(W^{1-\frac{1}{r},r}(\Gamma))$ imply that $D_{ij}^{\beta} \in L^r(W^{1-\frac{1}{r},r})$.

Therefore, the $W_r^{2,1}(Q_T)$-regularity is established.

The estimate (2.28) is then a direct consequence of an a priori estimate in Solonnikov [15], p. 162. ■

3 EXISTENCE IN $C^\gamma(W^{2,r}) \cap W^{\sigma,r}(W^{1,r}) \times C^\gamma(W^{1,r}) \cap W^{\sigma,r}(L^r) \times W_r^{2,1}(Q_T)$

After studying the auxiliary problems, we now turn to proving existence for the coupled system (2.1).

Theorem 3.1 *Let $0 \le \beta < 1$, $r > 2(n+1)$, and let the regularity assumptions (2.2) and the compatibility condition (2.27) hold. Further, suppose that there is a constant C_0, depending just on Ω and n such that,*

$$\left\|\frac{\partial f}{\partial s}\right\|_{L^\infty(\Omega\times R)} + \left\|\frac{\partial^2 f}{\partial s^2}\right\|_{L^\infty(\Omega\times R)} \le C_0\|\zeta'\|_{L^\infty(R)}, \tag{3.1}$$
$$\left\|\frac{\partial \lambda}{\partial s}\right\|_{L^\infty(\Omega\times R)} \le C_0C_2\|\zeta'\|_{L^\infty(R)},$$

and that

$$\|\zeta'\|_{L^\infty(R)} \le \Big[2C_2\max\{1, C_{emb}\}\cdot\Big\{1+\phi\Big(\overline{\psi}(\tfrac{1}{2})+\tfrac{1}{2}C_0\Big)\Big\}\cdot \cdot\{\|q^I\|_{L^r(L^r)} + \|c_0\|_{W^{2-\frac{2}{r},r}(\Omega)} + \|g\|_{W_r^{1-\frac{1}{r},\frac{1}{2}-\frac{1}{2r}}(S_T)}\}\Big]^{-1},$$

where the functions $\overline{\psi}$ and ϕ are defined by (3.2) and (2.28), respectively, and C_{emb} denotes the embedding constant for the embedding

$$W_r^{2,1}(Q_T) \hookrightarrow C^\gamma(W^{1,\infty}(\Omega)) \cap W^{\sigma,r}(L^r), \quad 1>\sigma>\frac{1}{2}+\frac{n+2}{2r}, \quad 0<\gamma<\frac{1}{2}-\frac{n+2}{2r}.$$

Then, (2.1) has at least one solution $\{p, v, c\}$ such that

$$\{p, v, c\} \in C^\gamma(W^{2,r}) \cap W^{\sigma,r}(W^{1,r}) \times C^\gamma(W^{1,r}) \cap W^{\sigma,r}(L^r) \times W_r^{2,1}(Q_T).$$

Proof: The assumption (3.1) allows us to simplify the estimates (2.20) and (2.22) for the velocity. By (3.1),

$$\widetilde{\psi}(y) \le \frac{C_3}{(1-y)^2}[A_3(q^I - q^P, f) + C_0y]. \tag{3.2}$$

Consequently, (2.20) and (2.22) turn out to be

$$\|v\|_{C^\gamma(W^{1,r})\cap W^{\sigma,r}(L^r)} \le \overline{\psi}(y), \tag{3.3}$$

where $y = C_2\|\zeta'\|_{L^\infty(R)}\|s\|_{C^\gamma(W^{1,\infty})}$ and $\widetilde{\psi}$ is given by

$$\overline{\psi}(y) = \{C_3A_3 + C_0y + C_D + C_E(1+C_0)y/C_2\}/(1-y)^2.$$

Now, let

$$s \in C^\gamma(W^{1,\infty}) \cap W^{\sigma,r}(L^r), \quad \frac{1}{2}+\frac{1}{r} < \gamma < 1-\frac{n+2}{r},$$

be such that $1 > \sigma > \frac{1}{2}+\frac{n+2}{2r}$. Then

$$\|s\|_{C^\gamma(W^{1,\infty})\cap W^{\sigma,r}(L^r)} < \max\{1, C_{emb}\}\Big\{\|q^I\|_{L^r(L^r)} + \|c_0\|_{W^{2-\frac{2}{r},r}(\Omega)} + \|g\|_{W_r^{1-\frac{1}{r},\frac{1}{2}-\frac{1}{2r}}(S_T)}\Big\}\phi\Big(\overline{\psi}\Big(\frac{1}{2}\Big)+\frac{1}{2}C_0\Big). \tag{3.4}$$

Next, we consider the problem

$$\begin{cases} v = -\zeta(s)K(x)[\nabla p - f(x,s)] & \text{in } Q_T, \\ \operatorname{div} v = q^I - q^P & \text{in } Q_T, \\ v \cdot \nu = 0 & \text{on } S_T, \\ \int_\Omega (q^I - q^P) = 0, \quad \int_\Omega p = 0; & \end{cases} \tag{3.5}$$

Lemmata 2.3 and 2.5 imply that (3.5) has a unique solution

$$\{p, v\} \in C^\gamma(W^{2,r}) \cap W^{\sigma,r}(W^{1,r}) \times C^\gamma(W^{1,r}) \cap W^{\sigma,r}(L^r), \quad r > 2(n+1), \quad \int_\Omega p = 0.$$

Inequalities (3.3) and (3.4) imply that

$$C_2 \|\zeta'\|_{L^\infty} \|s\|_{C^\gamma(W^{1,\infty})} \leq \frac{1}{2};$$

i.e., inequality (2.10) holds. Then, by using Lemmata 2.4 and 2.5 and inequality (3.3), we conclude that

$$\|v\|_{C^\gamma(W^{1,r}) \cap W^{\sigma,r}(L^r)} \leq \overline{\psi}\Big(\frac{1}{2}\Big). \tag{3.6}$$

Now, we consider the following problem:

$$\begin{cases} \dfrac{\partial c}{\partial t} - \operatorname{div}\{D^\beta(v,s)\nabla c - vc\} + q^P c = q^I & \text{in } Q_T, \\ D^\beta(v,s)\nabla c \cdot \nu = g & \text{on } S_T, \\ c(x,0) = c_0 & \text{in } \Omega, \\ D^\beta(u_0, c_0)\nabla c_0 \cdot \nu = g(x,0) & \text{in } \Omega, \end{cases} \tag{3.7}$$

where u_0 is the solution for (2.26). Lemma 2.6 implies the existence of a unique solution $c \in W_r^{2,1}(Q_T)$, $r > 2(n+1)$, for (3.7). Furthermore, the inequalities (2.28) and (3.6) imply that

$$\|c\|_{W_r^{2,1}(Q_T)} \leq \phi\Big(\overline{\psi}\Big(\frac{1}{2}\Big) + \frac{1}{2}C_0\Big)\Big\{\|q^I\|_{L^r(L^r)} + \|c_0\|_{W^{2-\frac{2}{r},r}(\Omega)} + \|g\|_{W_r^{1-\frac{1}{r},\frac{1}{2}-\frac{1}{2r}}(S_T)}\Big\}.$$

A result in Ladyženskaya *et al.* [8] implies that $\nabla c \in H^{\overline{\lambda},\frac{\overline{\lambda}}{2}}(\overline{Q}_T)$, where $0 \leq \overline{\lambda} < 1 - \frac{n+2}{r}$. Therefore, $W_r^{2,1}(Q_T)$ is compactly embedded in $C^\gamma(W^{1,\infty}(\Omega))$ for $\gamma < \frac{1}{2} - \frac{n+2}{2r}$, and let C_{emb} be the corresponding embedding constant.

Define a nonlinear operator T by setting

$$T(s) = c$$

and a Banach space B by setting

$$B = C^\gamma(W^{1,\infty}(\Omega)) \cap W^{\sigma,r}(L^r(\Omega)),$$

for $0 < \gamma < \frac{1}{2} - \frac{n+2}{2r}$ and $1 > \sigma > \frac{1}{2} + \frac{n+2}{2r}$.

Let a ball K_c be given by

$$K_c = \Big\{ z \in B : \|z\|_B \le \max\{1, C_{emb}\} \phi\Big(\overline{\psi}\Big(\frac{1}{2}\Big) + \frac{1}{2}C_0\Big) \cdot \\ \cdot \Big[\|c_0\|_{W^{2-\frac{2}{r},r}(\Omega)} + \|q^I\|_{L^r(L^r)} + \|g\|_{W_r^{1-\frac{1}{r},\frac{1}{r}-\frac{1}{2r}}(S_T)} \Big].$$

Then it is easy to see that T maps the ball K_c into itself. Obviously, K is a convex, closed, and bounded subset of B. Furthermore, $T(K_c)$ is a compact set in B. It remains to show that T is continuous. Then the Schauder fixed-point theorem will imply the statement of Theorem 3.1.

It is enough to show that T is sequentially continuous. Let $\{s_k\}$ be a sequence in K_c such that $s_k \to s$ in $C^\gamma(W^{1,\infty})$. Set $c_k = T(s_k)$ and denote solutions to (3.5) by $\{p_k, v_k\}$. By using estimates from Mikelić [9], p. 195, we see that

$$\begin{cases} \|p - p_k\|_{C(W^{2,r})} \le C_4 \|s - s_k\|_B, \\ \|v - v_k\|_{C(W^{2,r})} \le C_4 \|s - s_k\|_B, \end{cases} \tag{3.8}$$

where $\{p, v\}$ is a solution for (3.5) for the $s = \lim s_k$. Now, we consider (3.7) in the weak formulation, with a given velocity v_k:

$$\int_\Omega \frac{\partial c_k}{\partial t}\varphi + \int_\Omega D^\beta(v_k, s_k)\nabla c_k \nabla\varphi - \int_\Omega v_k c_k \nabla\varphi + \int_\Omega q^P c\varphi \\ = \int_\Omega q^I\varphi + \int_\Gamma g\varphi, \quad \forall\varphi \in H^1(\Omega), \quad \forall t \in [0, T].$$

Obviously, the sequence $\{c_k\}$ is uniformly bounded in $W_r^{2,1}(Q_T)$. Therefore, there exists a subsequence $\{c_{k_i}\}$ such that

$$\begin{cases} c_{k_i} \to c^* \text{ in } W_r^{2,1}(Q_T) \text{ weakly}, \\ c_{k_i} \to c^* \text{ in } C^\gamma(W^{1,\infty}) \text{ strongly}. \end{cases} \tag{3.9}$$

By using (3.8) and (3.9), we easily see that

$$\int_\Omega \frac{\partial c^*}{\partial t}\varphi + \int_\Omega D^\beta(v, s)\nabla c^* \nabla\varphi - \int_\Omega v c^* \nabla\varphi + \int_\Omega q^P c^*\varphi \\ = \int_\Omega q^P\varphi + \int_\Gamma g\varphi, \quad \forall\varphi \in H^1(\Omega), \quad \forall t \in [0, T].$$

Hence, $c^* \in W_r^{2,1}(Q_T)$ is a smooth weak solution for (3.7), and we conclude that it is also the strong solution for (3.7). Thus, $c^* = T(s) = c$. Uniqueness of the solution for (3.7) implies the convergence of the whole sequence $\{c_k\}$ towards $c = T(s)$. Therefore, $s_k \to s$ in B. ■

4 A UNIQUENESS RESULT

After proving existence of the smooth solution to (2.1), uniqueness is a corollary of Theorem 4, p. 1379, of Fabrie and Langlais [4].

Theorem 4.1 *Let the assumptions of Theorem 3.1 hold. Then, (2.1) has a unique solution in the class*

$$C^{\gamma}(W^{2,r}) \cap W^{\sigma,r}(W^{1,r}) \times C^{\gamma}(W^{1,r}) \cap W^{\sigma,r}(L^{r}) \times W_r^{2,r}(Q_T).$$

Proof: Obviously, velocity v is a bounded function and consequently D^{β} is Lipschitz continuous in v . Also, p and c are of higher smoothness than $L^2(W^{1,\infty}) \cap L^{\infty}(Q_T) \times L^{\infty}(W^{1,\infty})$. Hence, Theorem 4, p. 1379, of Fabrie and Langlais [4] implies the uniqueness. ∎

Acknowledgment. The author would like to thank Ms. Judy Stout for typing this difficult manuscript.

REFERENCES

[1] Antontsev S. N., Kazhikhov A. V., Monakhov V. *Boundary Value Problems in Mechanics of Non-Homogeneous Fluids.* North Holland, Amsterdam, 1990.

[2] Douglas J., Ewing R. E., Wheeler M. F. The approximation of the pressure by a mixed method in the simulation of miscible displacement. *R.A.I.R.O. Analyse Numerique*, 17:17–33, 1983.

[3] Ewing R. E., Wheeler M. F. Galerkin methods for miscible displacement problems in porous media. *SIAM J. Numer. Anal.*, 17:351–365, 1980.

[4] Fabrie P., Langlais M. Mathematical analysis of miscible displacement in porous medium. *SIAM J. Math. Analysis*, 23:1375–1392, 1992.

[5] Goyeau B. PhD thesis, Université de Bordeaux I, 1988.

[6] Grisvard P. *Elliptic Problems in Nonsmooth Domains.* Pitman, London, 1985.

[7] Hanouzet B. Applications bilinéaires compatibles avec un système à coefficients variables continuité dans les espaces de Besov. *Comm. in Partial Differential Equations*, 10:433–465, 1985.

[8] Ladyženskaya O. A, Solonnikov V. A., Ural'ceva N. N. *Linear and Quasilinear Equations of Parabolic Type. Translations of Mathematical Monographs 23*, American Mathematical Society, Providence, 1968.

[9] Mikelić A. Mathematical theory of stationary miscible filtration. *J. Differential Equations*, 90:186–202, 1991.

[10] Nečas J. *Les Méthodes Directes en Théorie des Équations Elliptiques.* Masson, Paris, 1967.

[11] Peaceman D. W. Improved treatment of dispersion in numerical calculation of multidimensional miscible treatment. *Soc. Pet. Eng. J.*, 6:213–216, 1966.

[12] Russell T. F., Wheeler M. F. Finite element and finite difference methods for continuous flow in porous media. In *The Mathematics of Reservoir Simulation*, pages 35–105, SIAM, Philadelphia, 1983. R. E. Ewing, ed.

[13] Russell T. F., Wheeler M. F., Chen C. Large scale simulation of miscible displacement by mixed and characteristic finite element methods. In *Mathematical and Computational Methods in Seismic Exploration and Reservoir Modeling*, pages 87–107, SIAM, Philadelphia, 1986. W. F. Fitzgibbon, ed.

[14] Sammon P. H. Numerical approximation for a miscible displacement process in porous media. *SIAM J. Numer. Anal.*, 23:507–542, 1986.

[15] Solonnikov V. A. Estimates in L_p of solutions of elliptic and parabolic systems. *Proc. Steklov Inst. Math.*, 102:157–185, 1967.

[16] Triebel H. *Interpolation Theory, Function Spaces, Differential Operators.* North Holland, Amsterdam, 1978.

International Series of Numerical Mathematics, Vol. 114, © 1993 Birkhäuser Verlag Basel

Distributed Microstructure Models of Porous Media

Ralph E. Showalter*

Abstract. Laminar flow through fissured or otherwise highly inhomogeneous media leads to very singular initial-boundary-value problems for equations with rapidly oscillating coefficients. The limiting case (by homogenization) is a continuous distribution of model cells which represent a valid approximation of the finite (singular) case, and we survey some recent results on the theory of such systems. This is developed as an application of continuous direct sums of Banach spaces which arise rather naturally as the energy or state spaces for the corresponding (stationary) variational or (temporal) dynamic problems. We discuss the basic models for a totally fissured medium, the extension to include secondary flux in partially fissured media, and the classical model systems which are realized as limiting cases of the microstructure models.

1 INTRODUCTION

Fractured or fissured porous media are commonly modelled as a composite material consisting of two components for which the internal flow is described by a pair of partial differential equations, one acting in each of the components, and a coupling that describes the interface between these components. For any region which consists of two finely interspersed materials, one can consider averaged properties of both materials as existing at every point in the region, and this leads to various classes of *double porosity* models. The classical example of this approach is the parabolic system

$$\begin{aligned} \frac{\partial}{\partial t}(au_1) - \vec{\nabla}\cdot(A\vec{\nabla}u_1) + \frac{1}{\delta}(u_1 - u_2) &= f_1, \\ \frac{\partial}{\partial t}(bu_2) - \vec{\nabla}\cdot(B\vec{\nabla}u_2) + \frac{1}{\delta}(u_2 - u_1) &= f_2, \end{aligned} \tag{1.1}$$

discussed in [7] for which u_1 represents the density of fluid in the first material and u_2 the density in the second. Similarly, $a(x)$ and $A(x)$ are porosity and permeability of the first material, respectively, while $b(x)$ and $B(x)$ are corresponding properties of the second material. The third term is an attempt to quantify the exchange of fluid between the two components. (See [38] for a corresponding heat conduction model.) Since the two components are treated symmetrically, such a double porosity model is said to be of *parallel flow* type. This symmetric treatment of the two components is a real limitation of these classical double porosity models. For a fractured medium, such a representation is particularly inappropriate, since the porous and permeable cells within the structure have flow properties radically different from those of the surrounding highly developed system of fissures. Moreover, the geometry of the individual cells and the corresponding

*Department of Mathematics, The University of Texas at Austin, Austin, Texas 78712-1082, USA

interface is lost in the averaging process leading to such models. Another class of double porosity models consists of the *distributed microstructure* models. At each point in the region there is given a representative model cell, the flow within each such local cell is described by an initial-boundary-value problem, and the boundary values on the cells are coupled to a single global initial-boundary-value problem which describes the global flow in the region. Thus we have a continuum of partial differential equations to describe the local flow on the micro-scale, and these are coupled to a single partial differential equation for the macro-scale flow. This concept occurred in a heat conduction problem in [31] and has arisen in a variety of applications which we mention below.

We shall illustrate these two classes of double porosity models in each of two models of fissured media. We use both types to describe first the *totally fissured* case in which the cells are individually and completely isolated from each other by the fissure system. In these models the cells act as storage sites only, and there is no direct diffusion from cell to cell, as they are connected only indirectly through the surrounding fissure system. Then we introduce corresponding models for the *partially fissured* case in which there is some fluid flow induced through the cell structure by the pressure gradient in the fissure system. This flow through the cells contributes an additional component to the velocity field in the fissure system which we call the *secondary flux*. Finally, we show the form of the functional differential equations that arise for the global flow when the local problems are eliminated from the system by Green's operator representations.

2 TOTALLY FISSURED MEDIA

A fractured medium consists of an ensemble of small porous and permeable cells which are surrounded by a highly developed system of fractures. The bulk of the flow occurs in the highly permeable fracture system, and most of the storage of fluid is in the system of cells which accounts for almost all of the total volume. One approach to constructing a model of such a medium is to regard the fissure system as the first component and the cell system as the second component of a *parallel flow* model obtained from (1.1) by adjusting the coefficients appropriately. In order to specialize the system (1.1) to a totally fissured medium in which the individual cells are isolated from each other and no direct cell-to-cell flow is possible, one sets $B = 0$. The resulting system of parabolic-ordinary type differential equations,

$$\frac{\partial}{\partial t}(au_1) - \vec{\nabla} \cdot (A\vec{\nabla}u_1) + \frac{1}{\delta}(u_1 - u_2) = f, \tag{2.1a}$$

$$\frac{\partial}{\partial t}(bu_2) + \frac{1}{\delta}(u_2 - u_1) = 0, \tag{2.1b}$$

is called the *first-order kinetic* model, since the cell storage is regarded as an added kinetic storage perturbation of the global fracture system. See [1], [13], [50], [11], [30], [21], [14], [9], [47], [29], [43] for applications and mathematical developments of such models.

Two essential limitations of the parallel-flow models are the suppression of the geometry of the cells and their corresponding interfaces on which the coupling occurs and the lack of any distinction between the space and time scales of the two components of the medium. The introduction of *distributed microstructure* models represents an attempt to recognize the geometry and the multiple scales in the problem as well as to better quantify the exchange of fluid across the intricate interface between the components. The global flow in the fracture system is described in the macro-scale x by

$$\frac{\partial}{\partial t}(a(x)u(x,t)) - \vec{\nabla} \cdot A(x)\vec{\nabla} u + q(x,t) = f(x,t), \qquad x \in \Omega, \tag{2.2a}$$

where $q(x,t)$ is the exchange term representing the flow into the cell Ω_x. The flow within the local cell Ω_x is described in the micro-scale variable y by

$$\frac{\partial}{\partial t}(b(x,y)U(x,y,t)) - \vec{\nabla}_y \cdot B(x,y)\vec{\nabla}_y U = F(x,y,t), \qquad y \in \Omega_x. \tag{2.2b}$$

Because of the smallness of the cells within the global region, the fissure pressure is assumed to be well approximated by the "constant" value $u(x,t)$ at every point of the cell boundary, so the effect of the fissures on the cell pressure is given by the interface condition

$$B(x,s)\vec{\nabla}_y U \cdot \nu + \frac{1}{\delta}(U - u) = 0, \qquad s \in \Gamma_s, \tag{2.2c}$$

where ν is the unit outward normal on Γ_x. (When $\delta = \infty$, this becomes (and converges to) the "matched" boundary condition, $u(x,t) = U(x,s,t)$ on Γ_x.) Finally, the amount of fluid flux across the interface scaled by the cell size determines the remaining term in (2.2a) by

$$q(x,t) = \frac{1}{|\Omega_x|}\int_{\Gamma_x} B(x,s)\vec{\nabla}_y U \cdot \nu \, ds, \tag{2.2d}$$

where $|\Omega_x|$ denotes the Lebesgue measure on Ω_x, and this contributes to the *cell storage*. Thus, the system (2.2) comprises a double-porosity model of "distributed microstructure" type for a totally fissured medium; it needs to be supplemented by appropriate boundary conditions for the global pressure $u(x,t)$ and initial conditions for $u(x,0)$ and $U(x,y,0)$ in order to comprise a well-posed problem. See [31], [37], [36], [15], [49], [8], [48], [6], [5], [4], [2], [15], [16], [17], [18], [20], [21], [22], [23], [19], [25], [26], [27], [24], [34], [46], [45], [44], [41], [42], [40] for applications and mathematical theory for (2.2) and various related problems.

Finally, we remark that the system (2.2) can be rewritten as a single equation of functional-differential type. By applying Gauss' theorem to (2.2b) we obtain from (2.2d)

$$\frac{\partial}{\partial t}\int_{\Omega_x} bU \, dy = \int_{\Gamma_x} B\frac{\partial U}{\partial \nu} \, ds + \int_{\Omega_x} F \, dy.$$

Use the Green's function for the problem (2.2b) to represent the solution $U(x,y,t)$ as an integral over Γ_x of $u(x,t)$ and substitute this in (2.2a) to get the implicit convolution evolution equation

$$\frac{\partial}{\partial t}\left\{a(x)u(x,t) + \int_0^t k(x,t-\tau)u(x,\tau)\, d\tau\right\} - \vec{\nabla} \cdot A(x)\vec{\nabla} u = f(x,t). \tag{2.3}$$

The convolution term represents a storage effect with memory. See [28] for a direct treatment and the very recent work of [35], where this equation forms the basis for an independent theoretical and numerical analysis.

3 PARTIALLY FISSURED MEDIA

Next we present a model for a partially fissured medium, that is, a fissured medium in which there are some flow paths directly joining the cells in addition to the predominate connection with the surrounding fissure system. Thus the cells are not completely isolated from one another by the fissure system. In this situation one must account for the effect of the gradient of the global flow on the local flow within the cells, for it is this fissure pressure gradient which necessarily provides the driving force for this cell-to-cell transport. In order to implement this in a model of parallel flow type, we introduce into the first order kinetic model (2.1) a secondary flux $\vec{u}_3$ through the cell system. This flux is assumed to respond to the fissure pressure gradient with a delay analogous to that of the cell storage in response to the value of fissure pressure in (2.1). The model system is then of the form

$$\frac{\partial}{\partial t}(au_1) - \vec{\nabla} \cdot (A\vec{\nabla} u_1) + \frac{1}{\delta}(u_1 - u_2) + \vec{\nabla} \cdot C^* \vec{u}_3 = f, \tag{3.1a}$$

$$\frac{\partial}{\partial t}(bu_2) + \frac{1}{\delta}(u_2 - u_1) = 0, \tag{3.1b}$$

$$\frac{\partial}{\partial t}(c\vec{u}_3) + \frac{1}{\beta}(\vec{u}_3 + C\vec{\nabla} u_1) = 0, \tag{3.1c}$$

where we assume the responses of the cell structure at a point to the value and to the gradient of fissure pressure are additive, an assumption that is valid for cell structures which are symmetric with respect to coordinate directions. The third and fourth terms in (3.1a) give the distributed mass flow rate into the cells from the fissure system at a point. According to (3.1b), the first of these goes toward the storage of fluid in the cells. Fluid from the fissure system enters the cell system at a point of higher pressure, it flows through the cell matrix to a point of lower pressure, and then it exits the cell back into the fissure system according to the second of these exchange terms in (3.1a). This results in a *secondary flux* $\vec{u}_3$ which follows the fissure pressure gradient according to (3.1c). The matrix C^*C arises from the bridging of the cells and it distinguishes the *partially fissured* model (3.1) from the *fully fissured* model (2.1). See [39] for a discussion and development of such models with multiple nonlinearities.

In order to obtain a *distributed microstructure* model of a partially fissured medium, we need to recognize that the cell system in (3.1) responds additively to the value and the gradient of fissure pressure, that is, to the best linear approximation of the fissure pressure at the cell location. As before, the global fluid flow in the fissures is described by

$$\frac{\partial}{\partial t}(a(x)u(x,t)) - \vec{\nabla} \cdot A(x)\vec{\nabla} u + q(x,t) = f(x,t), \qquad x \in \Omega, \tag{3.2a}$$

and the local flow in the cell at each point x is given by

$$\frac{\partial}{\partial t}(b(x,y)U(x,y,t)) - \vec{\nabla}_y \cdot B(x,y)\vec{\nabla}_y U = F(x,y,t), \qquad y \in \Omega_x. \tag{3.2b}$$

Our assumption that the cell pressure on the boundary is driven by the best linear approximation of the fissure pressure leads to the boundary condition

$$B(x,s)\vec{\nabla}_y U \cdot \nu + \frac{1}{\delta}(U - u - \vec{\nabla} u \cdot s) = 0, \qquad s \in \Gamma_s. \tag{3.2c}$$

Finally, the exchange term q in (3.2a) consists of two parts: the average amount flowing into the cell to be stored and the divergence of the secondary flux flowing through the cell structure. The total exchange is given by

$$q(x,t) = \frac{1}{|\Omega_x|}\int_{\Gamma_x} B(x,s)\vec{\nabla}_y U \cdot \nu\, ds - \frac{1}{|\Omega_x|}\vec{\nabla}\cdot\left(\int_{\Gamma_x} B(x,y)\vec{\nabla}_y U \cdot \nu s\, ds\right). \qquad (3.2d)$$

The system (3.2) comprises the distributed microstructure model for a partially fissured medium. This model was introduced in [10] to describe the highly anisotropic situation in layered media and developed in [12] for more general media. See [3] for a discrete version and numerical work.

When the cells Ω_x are symmetric in coordinate directions, one can separate the effects of storage from those of the secondary flux. Specifically, the storage can then be expressed in terms of the value of the fissure pressure at the point over a time interval through a convolution integral obtained as before from a Green's function representation of the cell problem, and the secondary flux and its corresponding contribution to the global flow are expressed likewise in terms of the global flux. This leads just as before to a functional partial differential equation of the form

$$\begin{aligned}&\frac{\partial}{\partial t}(a(x)u(x,t) + k_1(x,\cdot) * u(x,t))\\ &-\vec{\nabla}\cdot\left(A(x)\vec{\nabla}u(x,t) + \frac{\partial}{\partial t}k_{12}(x,\cdot) * \vec{\nabla}u(x,t) + k_2(x,\cdot) * \vec{\nabla}u(x,t)\right)\\ &= f(x,t), \qquad x\in\Omega,\ t>0.\end{aligned} \qquad (3.3)$$

which is known as *Nunziato's equation*. This equation was presented in [33] without any physical or philosophical justification as an interesting generalization of heat conduction with memory models due to Gurtin and Chen. See [32] for mathematical development of these equations.

4 REMARKS

The basic distributed microstructure model (2.2) is obtained as the limit by homogenization of a corresponding exact but highly singular partial differential equation with rapidly oscillating coefficients. This provides not merely another derivation of the model equations, but shows also the relation with the classical but singular case of a single diffusion equation, and it provides a method for directly computing the coefficients in (2.2) which necessarily represent averaged material properties. A similar result of convergence of a classical system to (3.2) is to be expected.

The first order kinetic model (2.1) is the limiting case of (2.2) as the permeability coefficient B tends to infinity. In this limit the cell behaves as a single point, equivalently, the function U is independent of the local variable y, and the geometry of the cell is lost. The system (3.2) converges to (3.1) likewise as B tends to infinity, and when δ tends to zero in (2.2) or (3.2), the limiting problem has the boundary conditions (2.2c) or (3.2c) replaced by the corresponding "matched" conditions of Dirichlet type.

An interesting open problem is the determination of the coefficients in the system (2.2) from measurements of data on the boundary of the global region. It would be particularly interesting to obtain information on the cell geometry from such boundary measurements.

The systems (1.1) and (2.1) comprise parabolic and degenerate parabolic dynamical systems, respectively, in the product space $L^2(\Omega) \times L^2(\Omega)$. The functional differential equations (2.3) and (3.3) lead to dynamical systems in $L^2(\Omega)$, but these are governed by C_0 semigroups without regularizing effects, and the estimates and techniques for these are comparatively difficult. These equations lack the parabolic structure one seeks in such models. However, the systems (2.2) and (3.2) do retain all of the parabolic structure and corresponding estimates and regularity of classical parabolic systems when they are posed on the spaces $L^2(\Omega) \times L^2(\Omega, L^2(\Omega_x))$.

Experience suggests that the distributed microstructure models are conceptually easy to work with, they provide accurate models which include the fine scales and geometry appropriate for many problems, and their theory can be developed in a straightforward manner using conventional techniques. The numerical analysis of these systems provides a natural application of parallel methods.

Acknowledgement. This material is based on work supported by a grant from the National Science Foundation.

REFERENCES

[1] Anzelius A. Über Erwärmung vermittels durchströmender Medien. *Zeit. Ang. Math. Mech.*, 6:291–294, 1926.

[2] Arbogast T. Gravitational forces in dual-porosity models of single phase flow. To appear.

[3] Arbogast T. The double porosity model for single phase flow in naturally fractured reservoirs. In *Numerical Simulation in Oil Recovery*, The IMA Volumes in Mathematics and its Applications 11, pages 23–45. Springer-Verlag, Berlin and New York, 1988. M. F. Wheeler, ed.

[4] Arbogast T. Analysis of the simulation of single phase flow through a naturally fractured reservoir. *SIAM J. Numer. Anal.*, 26:12–29, 1989.

[5] Arbogast T., Douglas J., Hornung U. Modeling of naturally fractured petroleum reservoirs by formal homogenization techniques. In *Frontiers in Pure and Applied Mathematics*, pages 1–19. Elsevier, Amsterdam, 1991. R. Dautray, ed.

[6] Arbogast T., Douglas J. Jr., Hornung U. Derivation of the double porosity model of single phase flow via homogenization theory. *SIAM J. of Math. Anal.*, 21:823–836, 1990.

[7] Barenblatt G. I., Zheltov I. P., Kochina I. N. Basic concepts in the theory of seepage of homogeneous liquids in fissured rocks. *J. Appl. Math. and Mech.*, 24:1286–1303, 1960.

[8] Barker J. A. Block-geometry functions characterizing transport in densely fissured media. *J. of Hydrology*, 77:263–279, 1985.

[9] Charlaix E., Hulin J. P., Plona T. J. Experimental study of tracer dispersion in sintered glass porous materials of variable compation. *Phys. Fluids*, 30:1690–1698, 1987.

[10] Clark G. W., Showalter R. E. Fluid flow in a layered medium. *Quarterly Appl. Math.* To appear.

[11] Coats K. H., Smith B. D. Dead-end pore volume and dispersion in porous media. *Trans. Soc. Petr. Eng.*, 231:73–84, 1964.

[12] Cook J. D., Showalter R. E. Microstructure diffusion models with secondary flux. Preprint.

[13] Deans H. A. A mathematical model for dispersion in the direction of flow in porous media. *Trans. Soc. Petr. Eng.*, 228:49–52, 1963.

[14] DiBenedetto E., Showalter R. E. A free-boundary problem for a degenerate parabolic system. *Jour. Diff. Eqn.*, 50:1–19, 1983.

[15] Diesler P. F. Jr., Wilhelm R. H. Diffusion in beds of porous solids: measurement by frequency response techniques. *Ind. Eng. Chem.*, 45:1219–1227, 1953.

[16] Douglas J., Paes Leme P. J., Arbogast T., Schmitt T. Simulation of flow in naturally fractured reservoirs. In *Proceedings, Ninth SPE Symposium on Reservoir Simulation*, pages 271–279, Dallas, Texas, 1987. Society of Petroleum Engineers. Paper SPE 16019.

[17] Friedman A., Knabner P. A transport model with micro- and macro-structure. Preprint.

[18] Friedman A., Tzavaras A. A quasilinear parabolic system arising in modelling of catalytic reactors. *Jour. Diff. Eqns.*, 70:167–196, 1987.

[19] Hornung U. Applications of the homogenization method to flow and transport through porous media. In *Flow and Transport in Porous Media*, Singapore. Summer School, Beijing 1988, World Scientific. Xiao Shutie, ed., to appear.

[20] Hornung U. Miscible displacement in porous media influenced by mobile and immobile water. In *Nonlinear Partial Differential Equations.* Springer, New York, 1988. P. Fife and P. Bates, eds.

[21] Hornung U. Homogenization of Miscible Displacement in Unsaturated Aggregated Soils. In *Progress in Nonlinear Differential Equations and Their Applications*, pages 143–153, Boston, 1991. Composite Media and Homogenization Theory, ICTP, Triest 1990, Birkhäuser. G. Dal Mase and G. F. Dell'Antonio, eds.

[22] Hornung U. Miscible displacement in porous media influenced by mobile and immobile water. *Rocky Mtn. Jour. Math.*, 21:645–669, 1991. Corr. pages 1153–1158.

[23] Hornung U. Modellierung von Stofftransport in aggregierten und geklüftten Böden. *Wasserwirtschaft*, 1, 1992. To appear.

[24] Hornung U., Jäger, W. Homogenization of reactive transport through porous media. In *EQUADIFF 1991*, Singapore. World Scientific Publishing. C. Perelló, ed., submitted 1992.

[25] Hornung U., Jäger W. A model for chemical reactions in porous media. In *Complex Chemical Reaction Systems. Mathematical Modeling and Simulation*, volume 47 of *Chemical Physics*, pages 318–334. Springer, Berlin, 1987. J. Warnatz and W. Jäger, eds.

[26] Hornung U., Jäger W. Diffusion, convection, adsorption, and reaction of chemicals in porous media. *J. Diff. Equations*, 92:199–225, 1991.

[27] Hornung U., Jäger W., Mikelić A. Reactive transport through an array of cells with semi-permeable membranes. In preparation.

[28] Hornung U., Showalter R. E. Diffusion models for fractured media. *Jour. Math. Anal. Appl.*, 147:69–80, 1990.

[29] Knabner P. Mathematishche modelle für den transport gelöster Stoffe in sorbierenden porösen medien. Technical Report 121, Univ. Augsburg, 1989.

[30] Lindstrom F. T., Narasimham M. N. L. Mathematical theory of a kinetic model for dispersion of previously distributed chemicals in a sorbing porous medium. *SIAM J. Appl. Math.*, 24:496–510, 1973.

[31] Lowan A. N. On the problem of the heat recuperator. *Phil. Mag.*, 17:914–933, 1934.

[32] Miller R. K. An integrodifferential equation for rigid heat conductors with memory. *J. Math. Anal. Appl.*, 66:313–332, 1978.

[33] Nunziato J. W. On heat conduction in materials with memory. *Quarterly Appl. Math.*, 29:187–204, 1971.

[34] Packer L., Showalter R. E. Distributed capacitance microstructure in conductors. To appear in Applicable Analysis.

[35] Peszenska M. *Mathematical analysis and numerical approach to flow through fissured media.* PhD thesis, Univ. Augsburg, May 1992.

[36] Rosen J. B. Kinetics of a fixed bed system for solid diffusion into spherical particles. *J. Chem. Phys.*, 20:387–394, 1952.

[37] Rosen J. B., Winshe W. E. The admittance concept in the kinetics of chromatography. *J. Chem. Phys.*, 18:1587–1592, 1950.

[38] Rubinstein L. I. On the problem of the process of propagation of heat in heterogeneous media. *Izv. Akad. Nauk SSSR, Ser. Geogr.*, 1, 1948.

[39] Rulla J., Showalter R. E. Diffusion in partially fissured media and implicit evolution equations. In preparation.

[40] Showalter R. E. Implicit evolution equations. In *Differential Equations and Applications, Vol. II*, pages 404–411, Columbus, OH, 1989. International Conf. on Theory and Applications of Differential Equations, University Press, Athens.

[41] Showalter R. E. Diffusion models with micro-structure. *Transport in Porous Media*, 6:567–580, 1991.

[42] Showalter R. E. Diffusion in a Fissured Medium with Micro-Structure. In *Free Boundary Problems in Fluid Flow with Applications*, volume 282 of *Pitman Research Notes in Mathematics*, pages 136–141. Longman, 1993. J. M. Chadam and H. Rasmussen, eds.

[43] Showalter R. E., Walkington N. J. A diffusion system for fluid in fractured media. *Differential and Integral Eqns.*, 3:219–236, 1990.

[44] Showalter R. E., Walkington N. J. Diffusion of fluid in a fissured medium with micro-structure. *SIAM Jour. Math. Anal.*, 22:1702–1722, 1991.

[45] Showalter R. E., Walkington N. J. Micro-structure models of diffusion in fissured media. *Jour. Math. Anal. Appl.*, 155:1–20, 1991.

[46] Showalter R. E., Walkington N. J. Elliptic systems for a medium with micro-structure. In *Geometric Inequalities and Convex Bodies*, pages 91–104. Marcel Dekker, New York, 1993. I. J. Bakelman, ed.

[47] van Duijn C. J., Knabner P. Solute transport through porous media with slow adsorption. To appear.

[48] van Genuchten M. Th., Wierenga P. J. Mass transfer studies in sorbing porous media I. Analytical Solutions. *Soil Sci. Soc. Amer. J.*, 40:473–480, 1976.

[49] Vogt Ch. A homogenization theorem leading to a Volterra integro-differential equation for permeation chromotography. Preprint #155, Sonderfachbereich 123, 1982.

[50] Warren J. E., Root P. J. The behavior of naturally fractured reservoirs. *Soc. Petr. Eng. J.*, 3:245–255, 1963.

International Series of Numerical Mathematics, Vol. 114, © 1993 Birkhäuser Verlag Basel

MULTIDIMENSIONAL DEGENERATE DIFFUSION PROBLEM WITH EVOLUTIONARY BOUNDARY CONDITION: EXISTENCE, UNIQUENESS, AND APPROXIMATION

N. Su (Su Ning)*

Abstract. This work studies an initial boundary value problem for nonlinear degenerate parabolic equations with evolutionary boundary conditions. Existence and uniqueness are established through some discrete schemes combined with parabolic regularization. Error estimates for these schemes in an L^2-sense are presented.

1 INTRODUCTION

This work is concerned with the nonlinear degenerate diffusion equation

$$\partial_t \theta(u) - \nabla \cdot a(\theta(u), \nabla u) = f(\theta(u)) \text{ in } (0,T) \times \Omega, \tag{1.1}$$

where $a(z,r)$ satisfies a certain ellipticity condition, $\theta(z)$ is nondecreasing, $f(z)$ is continuous, and $\Omega \subset \boldsymbol{R}^n$.

This type of equation occurs in a variety of physical situations and has been studied by many researchers (see, e.g., Alt and Luckhaus [1]). In the present paper, special attention is paid to the following boundary condition:

$$\partial_t \beta(u) + a(\theta(u), \nabla u) \cdot \nu = g(\beta(u)) \text{ on } (0,T) \times \Gamma, \tag{1.2}$$

where $\beta(z)$ is also nondecreasing, $g(z)$ is continuous, ν is the outward normal to $\partial\Omega$, and $\Gamma \subset \partial\Omega$.

The boundary condition (1.2) also appears in some physical settings. For instance, in the study of rainfall infiltration through soil, Darcy's law gives

$$-q = a(\theta, \nabla u) = K(\theta)(\nabla u + z), \tag{1.3}$$

where q represents the seepage velocity of the water, u the hydrostatic potential, $\theta = \theta(u)$ the moisture content, $K(\theta)$ the hydraulic conductivity, and z the gravitational potential. Let q_0 be the rate of rainfall. Then, on the surface of the soil,

$$-q \cdot \nu = q_0 \tag{1.4}$$

if the rainfall totally infiltrates into the soil. However, (1.4) fails to describe the partial infiltration. In practice, the water will accumulate on the ground surface as the surface

*Department of Applied Mathematics, Tsinghua University, Beijing, P. R. China

layer becomes saturated. The ponding rate will be $\partial_t u$, and thus (1.4) should be modified as follows:

$$H(u)\partial_t u - q \cdot \nu = q_0, \tag{1.5}$$

where $H(z) = 1$ if $z > 0$ and $H(z) = 0$ if $z < 0$. By setting $\beta(z) = \max\{z, 0\}$ we arrive at (1.2) with $g(z) = q_0$.

The present work studies a special case of (1.1), while the general situation will be investigated in a future paper.

Consider the problem

$$(I) \qquad \begin{cases} \partial_t\theta(u) - \nabla\cdot(\nabla u + b(\theta(u))) = f(\theta(u)) & \text{in } (0,T)\times\Omega, \\ \partial_t\beta(u) + (\nabla u + b(\theta(u)))\cdot\nu = g(\beta(u)) & \text{on } (0,T)\times\Gamma, \\ u = u_D & \text{on } (0,T)\times\Gamma', \\ \theta(u) = \theta_0 & \text{on } \{0\}\times\Omega, \\ \beta(u) = \beta_0 & \text{on } \{0\}\times\Gamma, \end{cases}$$

under the hypotheses:

(H_1) $\Omega \subset \boldsymbol{R}^n$ bounded, open, with Lipschitz boundary $\partial\Omega$; $\Gamma' \subset \partial\Omega$ measurable, with non-zero $(n-1)$-dimensional Hausdorff measure; $\Gamma = \partial\Omega \setminus \Gamma'$.

(H_2) The functions $\theta(z)$, $\beta(z)$, $b(z)$, $f(z)$ are uniformly Lipschitz continuous in $\boldsymbol{R}$ and, in addition,

$$\theta'(z) \geq 0, \quad \beta'(z) \geq 0, \ a.e. \text{ in } \boldsymbol{R}.$$

(H_3) $u_D \in H^1(0,T;H^1(\Omega)) \cap L^2(0,T;H^r(\Omega))$ with $r > 1$; there exists a function $u_0 \in H^r(\Omega)$ such that $\theta(u_0) = \theta_0$ in Ω and $\beta(u_0) = \beta_0$ on Γ.

By the weak solution of problem (I), we mean a function u such that

1° $u \in u_D + L^2(0,T;V), \theta(u) \in H^1(0,T;L^2(\Omega))$, $\beta(u) \in H^1(0,T;L^2(\Gamma))$, and

$$\theta(u)|_{t=0} = \theta_0, \qquad \beta(u)|_{t=0} = \beta_0.$$

2° For all $\varphi \in L^2(0,T;V)$ and $\tau \in (0,T)$,

$$\begin{aligned} \int_0^\tau (\partial_t\theta(u) \ &- \ f(\theta(u)), \varphi) + \int_0^\tau \langle \partial_t\beta(u) - g(\beta(u)), \varphi\rangle \\ &+ \ \int_0^\tau (\nabla_u + b(\theta(u)), \nabla\varphi) = 0, \end{aligned}$$

where $V = \{\varphi \in H^1(\Omega) : \varphi|_{\Gamma'} = 0\}$; $(\cdot,\cdot)$ and $\langle\cdot,\cdot\rangle$ represent the inner product, respectively, in the Hilbert spaces $L^2(\Omega)$ and $L^2(\Gamma)$.

In this paper, we plan to establish existence and uniqueness of weak solutions through a procedure of parabolic regularization combined with some discrete schemes and to present error estimates for these schemes. First, in §2 we introduce a family of parabolic regularized problems (I_ε) and prove that the solutions of (I_ε), if they exist, are convergent to a weak solution of (I) as $\varepsilon \to 0$. The rate of convergence is shown, and the uniqueness for a weak solution of (I) is proved in the same time. The existence problem for (I_ε) is solved in §3 by means of a time-discrete scheme. A time-continuous

Galerkin scheme is also discussed. Finally, in §4 we study a fully discrete scheme, obtained by combining two semi-discrete schemes appearing in the preceding section. An L^2-error estimate for the scheme is established.

It should be mentioned that the major technique used in this work is the same as used by Nochetto [5] and Verdi [6]. However, the problem we now consider contains an evolutionary boundary condition and possesses a different degeneration feature. In an earlier work, a similar problem in the one-dimensional case was studied by the author and a colleague with a different method (see Li and Su [4]).

2 EXISTENCE AND UNIQUENESS

In order to solve problem (I), we first consider its parabolic regularization:

$$(I_\varepsilon)\qquad \begin{cases} \partial_t\theta_\varepsilon(u) - \nabla\cdot(\nabla u + b(\theta_\varepsilon(u))) = f(\theta_\varepsilon(u)) & \text{in } (0,T)\times\Omega, \\ u = u_D & \text{on } (0,T)\times\Gamma', \\ \partial_t\beta_\varepsilon(u) + (\nabla u + b(\theta_\varepsilon(u)))\cdot\nu = g(\beta_\varepsilon(u)) & \text{on } (0,T)\times\Gamma, \\ u = u_0 & \text{on } \{0\}\times(\Omega\cup\Gamma), \end{cases}$$

where $\theta_\varepsilon(z) = \theta(z) + \varepsilon z$, $\beta_\varepsilon(z) = \beta(z) + \varepsilon z$. The weak solution of problem (I_ε) is defined in a similar manner to that of problem (I).

Lemma 2.1 *For every $\varepsilon > 0$, (I_ε) possesses a unique weak solution.*

It should be noted that a one-dimensional problem similar to (I_ε) was studied in Czou [3]. However, we have not found corresponding existence and uniqueness results for the general case. A proof of Lemma 2.1 is given in §3.

Now, suppose u_ε to be the weak solution of (I_ε). Then it is not hard to prove the following estimates, with the constants being independent of ε:

Lemma 2.2

$$\int_0^T \|\nabla u_\varepsilon\|^2_{L^2(\Omega)} \le C.$$

Lemma 2.3

$$\sup_{0\le t\le T} \|u_\varepsilon(t)\|^2_{H^1(\Omega)} + \int_0^T \left(\int_\Omega \theta'_\varepsilon(u_\varepsilon)|\partial_t u_\varepsilon|^2 + \int_\Gamma \beta'_\varepsilon(u_\varepsilon)|\partial_t u_\varepsilon|^2 \right) \le C.$$

Corollary 2.1

$$\int_0^T \|\partial_t\theta_\varepsilon(u_\varepsilon)\|^2_{L^2(\Omega)} + \int_0^T \|\partial_t\beta_\varepsilon(u_\varepsilon)\|^2_{L^2(\Gamma)} \le C,$$
$$\int_0^T \|\partial_t u_\varepsilon\|^2_{L^2(\Omega)} + \int_0^T \|\partial_t u_\varepsilon\|^2_{L^2(\Gamma)} \le C\varepsilon^{-1}.$$

When we consider error estimates for some discrete schemes related to (I_ε) in §§3 and 4, the following H^s-bounded for u_ε will be useful.

Lemma 2.4 *Let $1 < s < 3/2$, with $s \le r$ (see (H_3) in §1). Then,*

$$\sup_{0\le t\le T} \left\| \int_0^t u_\varepsilon \right\|^2_{H^s(\Omega)} + \int_0^T \|u_\varepsilon\|^2_{H^s(\Omega)} \le C.$$

Next, we discuss the convergence of the family of functions $\{u_\varepsilon\}$.

Lemma 2.5

$$\int_0^T \|\theta_\varepsilon(u_\varepsilon)-\theta_\delta(u_\delta)\|^2_{L^2(\Omega)}+\int_0^T \|\beta_\varepsilon(u_\varepsilon)-\beta_\delta(u_\delta)\|^2_{L^2(\Gamma)}$$
$$+\sup_{0\le t\le T}\left\|\nabla\int_0^t (u_\varepsilon-u_\delta)\right\|^2_{L^2(\Omega)}\le C(\varepsilon+\delta).$$

Proof: It follows from the definition of a weak solution that

$$\int_0^\tau (\partial_t(\theta_\varepsilon(u_\varepsilon))-\theta_\delta(u_\delta)),\varphi)+\int_0^\tau \langle\partial_t(\beta_\varepsilon(u_\varepsilon))-\beta_\delta(u_\delta)),\varphi\rangle$$
$$+\int_0^\tau (\nabla(u_\varepsilon-u_\delta),\nabla\varphi)+\int_0^\tau (b(\theta_\varepsilon(u_\varepsilon))-b(\theta_\delta(u_\delta)),\nabla\varphi)$$
$$=\int_0^\tau (f(\theta_\varepsilon(u_\varepsilon))-f(\theta_\delta(u_\delta)),\varphi)+\int_0^\tau \langle g(\beta_\varepsilon(u_\varepsilon))-g(\beta_\delta(u_\delta)),\varphi\rangle,$$

where $\tau\in(0,T)$ and $\varphi\in V$. Choose

$$\varphi(t)=\int_t^\tau (u_\varepsilon(s)-u_\delta(s))ds.$$

Then,

$$\int_0^\tau (\partial_t(\theta_\varepsilon(u_\varepsilon))-\theta_\delta(u_\delta)),\varphi)=(\theta_\varepsilon(u_\varepsilon))-\theta_\delta(u_\delta),\varphi)|_0^\tau-\int_0^\tau (\theta_\varepsilon(u_\varepsilon))-\theta_\delta(u_\delta),\partial_t\varphi)$$
$$=-(\theta_\varepsilon(u_0)-\theta_\delta(u_0),\varphi(0))+\int_0^\tau (\theta(u_\varepsilon)-\theta(u_\delta),u_\varepsilon-u_\delta)+\int_0^\tau (\varepsilon u_\varepsilon-\delta u_\delta,u_\varepsilon-u_\delta)$$
$$\ge\mu\int_0^\tau \|\theta(u_\varepsilon)-\theta(u_\delta)\|^2_{L^2(\Omega)}-\lambda\|\varphi(0)\|^2_{L^2(\Omega)}-C_\lambda(\varepsilon+\delta).$$

Similarly,

$$\begin{aligned}
\int_0^\tau \langle\partial_t(\beta_\varepsilon(u_\varepsilon)-\beta_\delta(u_\delta)),\varphi\rangle &\ge \mu\int_0^\tau \|\beta(u_\varepsilon)-\beta(u_\delta)\|^2_{L^2(\Gamma)}-\lambda\|\varphi(0)\|^2_{L^2(\Gamma)}-C_\lambda(\varepsilon+\delta),\\
\int_0^\tau \|\theta(u_\varepsilon)-\theta(u_\delta)\|^2_{L^2(\Omega)} &\ge \frac{1}{2}\int_0^\tau \|\theta_\varepsilon(u_\varepsilon)-\theta_\delta(u_\delta)\|^2_{L^2(\Omega)}-C(\varepsilon+\delta),\\
\int_0^\tau \|\beta(u_\varepsilon)-\beta(u_\delta)\|^2_{L^2(\Gamma)} &\ge \frac{1}{2}\int_0^\tau \|\beta_\varepsilon(u_\varepsilon)-\beta_\delta(u_\delta)\|^2_{L^2(\Gamma)}-C(\varepsilon+\delta),
\end{aligned}$$

since $\theta(z)=\theta_\varepsilon(z)-\varepsilon z$ and $\beta(z)=\beta_\varepsilon(z)-\varepsilon z$ for every $\varepsilon>0$. Moreover,

$$\begin{aligned}
\int_0^\tau (\nabla(u_\varepsilon-u_\delta),\nabla\varphi) &= -\int_0^\tau (\partial_t\nabla\varphi,\nabla\varphi)=\frac{1}{2}\|\nabla\varphi(0)\|^2_{L^2(\Omega)},\\
\left|\int_0^\tau (b(\theta_\varepsilon(u_\varepsilon))-b(\theta_\delta(u_\delta)),\nabla\varphi)\right| &\le \lambda\int_0^\tau \|\theta_\varepsilon(u_\varepsilon)-\theta_\delta(u_\delta)\|^2_{L^2(\Omega)}+C_\lambda\int_0^\tau \|\nabla\varphi\|^2_{L^2(\Omega)},\\
\int_0^\tau (f(\theta_\varepsilon(u_\varepsilon))-f(\theta_\delta(u_\delta)),\varphi) &\le \lambda\int_0^\tau \|\theta_\varepsilon(u_\varepsilon)-\theta_\delta(u_\delta)\|^2_{L^2(\Omega)}+C_\lambda\int_0^\tau \|\varphi\|^2_{L^2(\Omega)},\\
\int_0^\tau \langle g(\beta_\varepsilon(u_\varepsilon))-g(\beta_\delta(u_\delta)),\varphi\rangle &\le \lambda\int_0^\tau \|\beta_\varepsilon(u_\varepsilon)-\beta_\delta(u_\delta)\|^2_{L^2(\Gamma)}+C_\lambda\int_0^\tau \|\varphi\|^2_{L^2(\Gamma)}.
\end{aligned}$$

Combining the above and choosing $\lambda > 0$ suitably small, we obtain

$$\int_0^T \|\theta_\varepsilon(u_\varepsilon) - \theta_\delta(u_\delta)\|^2_{L^2(\Omega)} + \int_0^T \|\beta_\varepsilon(u_\varepsilon) - \beta_\delta(u_\delta)\|^2_{L^2(\Gamma)} + \left\|\nabla \int_0^T (u_\varepsilon - u_\delta)\right\|^2_{L^2(\Omega)} \leq C\left(\varepsilon + \delta + \int_0^T \left\|\nabla \int_t^T (u_\varepsilon - u_\delta)\right\|^2_{L^2(\Omega)}\right);$$

the embedding inequality

$$\|\varphi\|_{L^2(\Omega)} + \|\varphi\|_{L^2(\Gamma)} \leq C\|\nabla\varphi\|_{L^2(\Omega)}$$

was used.

An argument of Gronwall type yields the conclusion of the lemma. ■

From Lemma 2.5, Lemma 2.2, and Corollary 2.1, we can easily prove the following theorem.

Theorem 2.1 *Problem* (I) *possesses a unique weak solution u. Furthermore, the weak solution u satisfies*

$$\int_0^T \|\theta_\varepsilon(u_\varepsilon) - \theta(u)\|^2_{L^2(\Omega)} + \int_0^T \|\beta_\varepsilon(u_\varepsilon) - \beta(u)\|^2_{L^2(\Gamma)} + \sup_{0\leq t\leq T} \left\|\nabla \int_0^t (u_\varepsilon - u)\right\|^2_{L^2(\Omega)} \leq C_\varepsilon.$$

Note that the proof of Lemma 2.5 implies the uniqueness of (I_ε) as well as that of (I).

3 SEMI-DISCRETE SCHEMES

In this section, we consider two semi-discrete schemes. Each can be used to prove the existence of weak solutions of (I_ε).

To simplify notation, we omit the regularization parameter ε everywhere in this section; that is, we set $\theta(z) = \theta_\varepsilon(z)$, $\beta(z) = \beta_\varepsilon(z)$, and $u = u_\varepsilon$. This should not cause confusion since the original problem (I) is not used in this section.

3.1 Discrete time scheme

We replace the time derivative by the backward difference quotient in (I_ε). Let N be a positive integer and $\tau = T/N$; denote by $\{U^j\}_{j=0,1,\ldots,N}$ the solution of the elliptic system

$$(I_\varepsilon)^\tau \quad \begin{cases} \Big(\frac{1}{\tau}(\theta(U^j) - \theta(U^{j-1})) - f(\theta(U^j)), \varphi) + (\nabla U^j + b\theta(U^j)), \nabla\varphi\Big) & \\ \qquad + \langle\frac{1}{\tau}(\beta(U^j) - \beta(U^{j-1})) - g(\beta(U^j)), \varphi\rangle = 0, & \forall \varphi \in V, \\ U^j \in \tilde{u}_D^j + V, & j = 1, 2, \ldots, N, \\ U^0 = u_0, & \end{cases}$$

where

$$\widetilde{u}_D^j = \frac{1}{\tau}\int_{(j-1)\tau}^{j\tau} u_D(t)dt.$$

We see from the weak solution of (I_ε) that

$$(\widetilde{I}_\varepsilon)^\tau \qquad \begin{cases} \left(\frac{1}{\tau}(\theta(u^j) - \theta(u^{j-1})) - \widetilde{f}^j, \varphi\right) + (\nabla \widetilde{u}^j + \widetilde{b}^j, \nabla\varphi) & \\ \qquad + \left\langle \frac{1}{\tau}(\beta(u^j) - \beta(u^{j-1})) - \widetilde{g}^j, \varphi\right\rangle = 0, & \forall \varphi \in V, \\ \widetilde{u}^j \in \widetilde{u}_D^j + V, & j = 1, 2, \ldots, N, \\ u^0 = u_0, & \end{cases}$$

where

$$u^j = u(j\tau), \widetilde{u}^j = \frac{1}{\tau}\int_{(j-1)\tau}^{j\tau} u(t)dt, \quad \widetilde{g}^j = \frac{1}{\tau}\int_{(j-1)\tau}^{j\tau} g(\beta(u(t)))dt,$$
$$\widetilde{b}^j = \frac{1}{\tau}\int_{(j-1)\tau}^{j\tau} b(\theta(u(t)))dt, \qquad \widetilde{f}^j = \frac{1}{\tau}\int_{(j-1)\tau}^{j\tau} f(\theta(u(t)))dt.$$

In order to estimate the error for $(I_\varepsilon)^\tau$, we subtract $(\widetilde{I}_\varepsilon)^\tau$ from $(I_\varepsilon)^\tau$ and obtain the relation

$$\begin{aligned} &\left(\frac{1}{\tau}(\theta(U^j) - \theta(u^j)) - \frac{1}{\tau}(\theta(U^{j-1}) - \theta(u^{j-1})), \varphi\right) + (\nabla(U^j - \widetilde{u}^j), \nabla\varphi) \\ &\quad + \left\langle \frac{1}{\tau}(\beta(U^j) - \beta(u^j)) - \frac{1}{\tau}(\beta(U^{j-1}) - \beta(u^{j-1})), \varphi\right\rangle \\ &= (f(\theta(U^j)) - \widetilde{f}^j, \varphi) - (b(\theta(U^j)) - \widetilde{b}^j, \nabla\varphi) \\ &\quad + \langle g(\beta(U^j)) - \widetilde{g}^j, \varphi\rangle, \qquad \forall \varphi \in V. \end{aligned} \tag{3.1}$$

Then we choose $\varphi = \varphi^j = \sum_{i=j}^m (U^i - \widetilde{u}^i)$, where $1 \le m \le N$, and sum with respect to j from 1 through m. We estimate its terms separately.

For the first term on the left hand side, by using the summation formula

$$\sum_{j=1}^m (a^j - a^{j-1}) \sum_{i=j}^m c^i = -a^0 \sum_{j=1}^m c^j + \sum_{j=1}^m a^j c^j, \tag{3.2}$$

we have

$$\begin{aligned} I_1 &= \sum_{j=1}^m \left(\frac{1}{\tau}(\theta(U^j) - \theta(u^j)) - \frac{1}{\tau}(\theta(U^{j-1}) - \theta(u^{j-1})), \sum_{i=j}^m (U^i - \widetilde{u}^i)\right) \\ &= -\frac{1}{\tau}\left(\theta(U^0) - \theta(u^0), \sum_{j=1}^m (U^j - \widetilde{u}^j)\right) + \frac{1}{\tau}\sum_{j=1}^m (\theta(U^j) - \theta(u^j), U^j - \widetilde{u}^j) \\ &= \frac{1}{\tau}\sum_{j=1}^m (\theta(U^j) - \theta(u^j), U^j - u^j) + \frac{1}{\tau}\sum_{j=1}^m (\theta(U^j) - \theta(u^j), u^j - \widetilde{u}^j). \end{aligned}$$

Since $0 \le \theta'(z) \le C$, Corollary 2.1 implies that, for some $\mu > 0$,

$$I_1 \ge \mu \frac{1}{\tau}\sum_{j=1}^m \|\theta(U^j) - \theta(u^j)\|_{L^2(\Omega)}^2 - C\varepsilon^{-1}.$$

In the same way,

$$\begin{aligned} I_2 &= \sum_{j=1}^{m}\Big\langle \frac{1}{\tau}(\beta(U^j)-\beta(u^j)) - \frac{1}{\tau}(\beta(U^{j-1})-\beta(u^{j-1})), \sum_{i=1}^{m}(U^i-\widetilde{u}^i)\Big\rangle \\ &\geq \mu\frac{1}{\tau}\sum_{j=1}^{m}\|\beta(U^j)-\beta(u^j)\|^2_{L^2(\Gamma)} - C\varepsilon^{-1}. \end{aligned}$$

The third term is positive:

$$I_3 = \sum_{j=1}^{m}\Big(\nabla(U^j-\widetilde{u}^j), \sum_{i=j}^{m}\nabla(U^i-\widetilde{u}^i)\Big) \geq \frac{1}{2}\Big\|\sum_{j=1}^{m}\nabla(U^j-\widetilde{u}^j)\Big\|^2_{L^2(\Omega)}.$$

The terms on the right hand side are treated as follows. Since

$$|f(\theta(U^j)) - \widetilde{f}^j| = \Big|f(\theta(U^j)) - f(\theta(u^j)) + \frac{1}{\tau}\int_{(j-1)\tau}^{j\tau}(f(\theta(u^j)) - f(\theta(u)))dt\Big|,$$

another application of Corollary 2.1 gives

$$\begin{aligned} I_4 &= \sum_{j=1}^{m}\Big(f(\theta(U^j)) - \widetilde{f}^j, \sum_{i=j}^{m}(U^i-\widetilde{u}^i)\Big) \\ &\leq \lambda\frac{1}{\tau}\sum_{j=1}^{m}\|\theta(U^j)-\theta(u^j)\|^2_{L^2(\Omega)} + C_\lambda\Big(1+\tau\sum_{j=1}^{m}\Big\|\sum_{i=j}^{m}(U^i-\widetilde{u}^i)\Big\|^2_{L^2(\Omega)}\Big). \end{aligned}$$

Similarly, we have

$$\begin{aligned} I_5 &= -\sum_{j=1}^{m}\Big(b(\theta(U^j)) - \widetilde{b}^j, \sum_{i=j}^{m}\nabla(U^i-\widetilde{u}^i)\Big) \\ &\leq \lambda\frac{1}{\tau}\sum_{j=1}^{m}\|\theta(U^j)-\theta(u^j)\|^2_{L^2(\Omega)} + C_\lambda\Big(1+\tau\sum_{j=1}^{m}\Big\|\sum_{i=j}^{m}\nabla(U^i-\widetilde{u}^i)\Big\|^2_{L^2(\Omega)}\Big), \\ I_6 &= \sum_{j=1}^{m}\Big\langle g(\beta(U^j)) - \widetilde{g}^j, \sum_{i=j}^{m}(U^i-\widetilde{u}^i)\Big\rangle \\ &\leq \lambda\frac{1}{\tau}\sum_{j=1}^{m}\|\beta(U^j)-\beta(u^j)\|^2_{L^2(\Gamma)} + C_\lambda\Big(1+\tau\sum_{j=1}^{m}\Big\|\sum_{i=j}^{m}(U^i-\widetilde{u}^i)\Big\|^2_{L^2(\Gamma)}\Big). \end{aligned}$$

Finally, by substituting the above estimates into (3.1) and using embedding inequalities, we obtain

$$\begin{aligned} \tau\sum_{j=1}^{m}\|\theta(U^j)-\theta(u^j)\|^2_{L^2(\Omega)} &+ \tau\sum_{j=1}^{m}\|\beta(U^j)-\beta(u^j)\|^2_{L^2(\Gamma)} + \Big\|\tau\sum_{j=1}^{m}\nabla(U^j-\widetilde{u}^j)\Big\|^2_{L^2(\Omega)} \\ &\leq C\Big(\varepsilon^{-1}\tau^2 + \tau\sum_{j=1}^{m}\Big\|\tau\sum_{i=j}^{m}\nabla(U^i-\widetilde{u}^i)\Big\|^2_{L^2(\Omega)}\Big). \end{aligned}$$

By using a discrete form of the Gronwall argument, we complete the proof of the following error estimate.

Theorem 3.1

$$\tau\sum_{j=1}^{N}\|\theta(U^j)-\theta(u^j)\|^2_{L^2(\Omega)} + \tau\sum_{j=1}^{N}\|\beta(U^j)-\beta(u^j)\|^2_{L^2(\Gamma)} + \max_{0\le m\le N}\left\|\nabla\Big(\tau\sum_{j=1}^{m}U^j-\int_0^{m\tau}u(t)dt\Big)\right\|^2_{L^2(\Omega)} \le C\varepsilon^{-1}\tau^2.$$

Employing the solution $\{U^j\}$ of $(I_\varepsilon)^\tau$ we outline here a proof of Lemma 2.1. As in the proofs of Lemmata 2.2 and 2.3, we can show that

$$\tau\sum_{j=1}^{N}\|\nabla U^j\|^2_{L^2(\Omega)}\le C, \tag{3.3}$$

$$\tau\sum_{j=1}^{N-m}\left\|\frac{(\theta(U^{m+j})-\theta(U^j))}{m\tau}\right\|^2_{L^2(\Omega)} + \tau\sum_{j=1}^{N-m}\left\|\frac{(\beta(U^{m+j})-\beta(U^j))}{m\tau}\right\|^2_{L^2(\Gamma)}\le C, \tag{3.4}$$

where $1\le m\le N$ is arbitrary. Then define the function $u^\tau:[0,T]\to H^1(\Omega)$ as follows:

$$u^\tau(t)=U^j \text{ if } t\in((j-1)\tau, j\tau].$$

From (3.3) and (3.4), it follows that there exists a subsequence $\{u^{\tau_k}\}$ and a function u such that, as $\tau_k\to 0$,

$$u^{\tau_k}\to u \quad \text{in} \quad L^2((0,T)\times\Omega),$$
$$u^{\tau_k}|_\Gamma\to u|_\Gamma \quad \text{in} \quad L^2((0,T)\times\Gamma).$$

Also, from (3.4), we see that $\theta(u)\in H^1(0,T;L^2(\Omega))$ and $\beta(u)\in H^1(0,T;L^2(\Gamma))$. From (3.3) we can assume that

$$u^{\tau_k}\rightharpoonup u \text{ in } L^2(0,T;H^1(\Omega)) \text{ as } \tau_k\to 0.$$

Finally, it is not hard to verify that u is a weak solution of (I_ε).

The uniqueness of the solution of (I_ε), as we have noted, can be proved in the same way as in the proof of Lemma 2.5.

3.2 Galerkin-finite element scheme

To study this scheme we make an additional hypothesis:

$(H_1)'$ Ω is a convex polygonal domain and satisfies (H_1).

Let V_h be a quasi-regular finite element space on Ω (see Ciarlet [2], for the details), and set $V_h^0=V\cap V_h$. Consider the solution $u_h(t)$ of the following problem:

$$(I_\varepsilon)_h\quad\begin{cases}(\partial_t\theta(u_h)-f(\theta(u_h)),\varphi)+(\nabla u_h+b(\theta(u_h)),\nabla\varphi)\\ \qquad +\langle\partial_t\beta(u_h)-g(\beta(u_h)),\varphi\rangle=0, & \forall\varphi\in V_h^0,\\ u_h(t)\in u_{Dh}(t)+V_h^0, & t\in(0,T),\\ u_h(0)=u_{0h},\end{cases}$$

where $u_{Dh} \in L^2(0,T;V_h)$ and $u_{0h} \in V_h$ satisfy the inequalities

$$\begin{aligned} \int_0^T \|u_{Dh} - u_D\|^2_{L^2(\Omega)} + h^2 \int_0^T \|\nabla(u_{Dh} - u_D)\|^2_{L^2(\Omega)} &\le \Omega_D(h), \\ \|\theta(u_{0h}) - \theta(u_0)\|^2_{L^2(\Omega)} + \|\beta(u_{0h}) - \beta(u_0)\|^2_{L^2(\Gamma)} &\le \Omega_0(h), \end{aligned}$$

respectively, with $\Omega_D(h) \to 0$, $\Omega_0(h) \to 0$ as $h \to 0$.

Let u be the weak solution of (I_ε). We first study its H^1-projection $\widehat{u}$ on V_h:

$$(P)_h \quad \begin{cases} (\nabla(u - \widehat{u}), \nabla\varphi) = 0, & \forall\varphi \in V_h^0, \\ \widehat{u}(t) \in u_{Dh}(t) + V_h^0. \end{cases}$$

From standard estimates for finite element approximations and elliptic equations, we have the following estimate.

Lemma 3.1 *Let s be as in Lemma 2.4. Then,*

$$\int_0^T \|u - \widehat{u}\|^2_{L^2(\Omega)} + h^2 \int_0^T \|\nabla(u - \widehat{u})\|^2_{L^2(\Omega)} \le C(h^{2s} + \Omega_D(h)), \tag{3.5}$$

$$\sup_{0\le t\le T} \left\| \nabla \int_0^t (u - \widehat{u}) \right\|^2_{L^2(\Omega)} \le Ch^{-2}(h^{2s} + \Omega_D(h)), \tag{3.6}$$

$$\int_0^T \|u - \widehat{u}\|^2_{L^2(\Gamma)} \le Ch^{-1}(h^{2s} + \Omega_D(h)). \tag{3.7}$$

In what follows, we perform an error estimate for scheme $(I_\varepsilon)_h$. By combining (I_ε), $(I_\varepsilon)_h$, and $(P)_h$, we see that

$$\begin{aligned} &\int_0^T (\partial_t(\theta(u) - \theta(u_h)), \varphi) + \int_0^T (\nabla(\widehat{u} - u_h), \nabla\varphi) + \int_0^T \langle \partial_t(\beta(u) - \beta(u_h)), \varphi\rangle \\ &\quad = \int_0^T (f(\theta(u)) - f(\theta(u_h)), \varphi) - \int_0^T (b(\theta(u)) - b(\theta(u_h)), \nabla\varphi) \\ &\qquad + \int_0^T \langle g(\beta(u)) - g(\beta(u_h)), \varphi\rangle, \qquad \forall\varphi \in V_h^0. \end{aligned}$$

Take $\varphi(t) = -\int_T^t (\widehat{u}(s) - u_h(s))ds$ and treat the terms one at a time. Then,

$$\begin{aligned} II_1 &= \int_0^T (\partial_t(\theta(u) - \theta(u_h)), \varphi) \\ &= -(\theta(u_0) - \theta(u_{0h}), \varphi(0)) + \int_0^T (\theta(u) - \theta(u_h), \widehat{u} - u_h) \\ &\ge -\lambda\|\varphi(0)\|^2_{L^2(\Omega)} - C_\lambda\Omega_0(h) + \int_0^T (\theta(u) - \theta(u_h), u - u_h + \widehat{u} - u). \end{aligned}$$

Since $\theta'(z) \le \mu^{-1}$, Lemma 3.1 indicates that

$$\left\| \int_0^T (\theta(u) - \theta(u_h), \widehat{u} - u) \right\| \le \lambda \int_0^T \|\theta(u) - \theta(u_h)\|^2_{L^2(\Omega)} + C_\lambda(h^{2s} + \Omega_D(h));$$

thus,

$$II_1 \ge (\mu - \lambda) \int_0^T \|\theta(u) - \theta(u_h)\|^2_{L^2(\Omega)} - \lambda\|\varphi(0)\|^2_{L^2(\Omega)} - C_\lambda(h^{2s} + \Omega_D(h) + \Omega_0(h)).$$

Analogously,

$$\begin{aligned} II_2 &= \int_0^T \langle \partial_t(\beta(u) - \beta(u_h)), \varphi \rangle \\ &\geq (\mu - \lambda) \int_0^T \|\beta(u) - \beta(u_h)\|^2_{L^2(\Gamma)} - \lambda \|\varphi(0)\|^2_{L^2(\Gamma)} \\ &\quad - C_\lambda (h^{2s-1} + h^{-1}\Omega_D(h) + \Omega_0(h)). \end{aligned}$$

Moreover,

$$\begin{aligned} II_3 &= \int_0^T (\nabla(\widehat{u} - u_h), \nabla\varphi) = \frac{1}{2}\|\nabla\varphi(0)\|^2_{L^2(\Omega)}, \\ II_4 &= \int_0^T (f(\theta(u)) - f(\theta(u_h)), \varphi) \\ &\leq \lambda \int_0^T \|\theta(u) - \theta(u_h)\|^2_{L^2(\Omega)} + C_\lambda \int_0^T \|\varphi\|^2_{L^2(\Omega)}, \\ II_5 &= -\int_0^T (b(\theta(u)) - b(\theta(u_h)), \nabla\varphi) \\ &\leq \lambda \int_0^T \|\theta(u) - \theta(u_h)\|^2_{L^2(\Omega)} + C_\lambda \int_0^T \|\nabla\varphi\|^2_{L^2(\Omega)}, \\ II_6 &= \int_0^T \langle g(\beta(u)) - g(\beta(u_h)), \varphi \rangle \\ &\leq \lambda \int_0^T \|\beta(u) - \beta(u_h)\|^2_{L^2(\Gamma)} + C_\lambda \int_0^T \|\varphi\|^2_{L^2(\Gamma)}. \end{aligned}$$

But it is known that $II_1 + II_2 + II_3 = II_4 + II_5 + II_6$. Therefore,

$$\begin{aligned} &\int_0^T \|\theta(u) - \theta(u_h)\|^2_{L^2(\Omega)} + \int_0^T \|\beta(u) - \beta(u_h)\|^2_{L^2(\Gamma)} + \left\|\nabla \int_0^T (\widehat{u} - u_h)\right\|^2_{L^2(\Omega)} \\ &\leq C\Big(h^{-1}(h^{2s} + \Omega_D(h)) + \Omega_0(h) + \int_0^T \left\|\nabla \int_t^T (\widehat{u} - u_h)\right\|^2_{L^2(\Omega)}\Big). \end{aligned}$$

Consequently, a Gronwall argument and Lemma 3.1 imply the following result.

Theorem 3.2 *Let s be as in Lemma 2.4. Then,*

$$\begin{aligned} &\int_0^T \|\theta(u) - \theta(u_h)\|^2_{L^2(\Omega)} + \int_0^T \|\beta(u) - \beta(u_h)\|^2_{L^2(\Gamma)} + \sup_{0\leq t\leq T} \left\|\nabla \int_0^t (\widehat{u} - u_h)\right\|^2_{L^2(\Omega)} \\ &\leq C\{h^{-1}(h^{2s} + \Omega_D(h)) + \Omega_0(h)\} \sup_{0\leq t\leq T} \left\|\nabla \int_0^t (u - u_h)\right\|^2_{L^2(\Omega)} \\ &\leq C\{h^{-2}(h^{2s} + \Omega_D(h)) + \Omega_0(h)\}. \end{aligned}$$

From the proof of Theorem 3.2, we see that, if $\beta(z) \equiv 0$,

$$\int_0^T \|\theta(u) - \theta(u_h)\|^2_{L^2(\Omega)} + \sup_{0\leq t\leq T} \left\|\nabla \int_0^t (\widehat{u} - u_h)\right\|^2_{L^2(\Omega)} \leq C(h^{2s} + \Omega_D(h) + \Omega_0(h)).$$

It can also be seen that the hypothesis $(H_1)'$ can be replaced by

$(H_1)''$ Ω satisfies (H_1) and $\partial\Omega \in C^2$.

4 DISCRETE SCHEME

The regularization parameter ε is still omitted. Consider the following scheme:

$$(I_\varepsilon)_h^\tau \quad \begin{cases} \Big(\frac{1}{\tau}(\theta(U_h^j) - \theta(U_h^{j-1})) - f(\theta(U^j)), \varphi\Big) + \nabla U_h^j + b(\theta(U_h^j)), \nabla\varphi) \\ \qquad +\Big\langle\frac{1}{\tau}(\beta(U_h^j) - \beta(U_h^{j-1})) - g(\beta(U_h^j)), \varphi\Big\rangle = 0, \qquad \forall\varphi \in V_h^0, \\ U_h^j \in \widetilde{u}_{Dh}^j + V_h^0, \qquad j = 1, 2, \ldots, N, \qquad (\tau = T/N), \\ U_h^0 = u_{0h}, \end{cases}$$

where $\widetilde{u}_{Dh}^j = \frac{1}{\tau}\int_{(j-1)\tau}^{j\tau} u_{Dh}(s)ds$ and u_{Dh}, u_{0h}, and V_h^0 are the same as in §3.

An error estimate can be proved in almost the same way as was used in the preceding sections. To avoid repetition, we will give only a sketch of the proof.

Let $\{U_h^j\}_{j=0,1,\ldots,N}$ be the solution of $(I_\varepsilon)_h^\tau$, $u = u_\varepsilon$ the weak solution of (I_ε), and $\widehat{u}$ the H^1-projection of u on V_h, as given by $(P)_h$ in §3. Set

$$\widetilde{u}^j = \frac{1}{\tau}\int_{(j-1)\tau}^{j\tau} u(s)ds, \quad \widehat{u}^j = \frac{1}{\tau}\int_{(j-1)\tau}^{j\tau} \widehat{u}(s)ds.$$

Then, from $(P)_h$ it follows that

$$(\widetilde{P})_h \quad \begin{cases} (\nabla(\widetilde{u}^j - \widehat{u}^j), \nabla\varphi) = 0, & \forall\varphi \in V_h^0, \\ \widehat{u}^j \in \widetilde{u}_{Dh}^j + V_h^0, & j = 1, 2, \ldots, N. \end{cases}$$

The following lemma is analogous to Lemma 3.1.

Lemma 4.1

$$\begin{aligned} h^2\tau\sum_{j=1}^N \|\nabla(\widetilde{u}^j - \widehat{u}^j)\|_{L^2(\Omega)}^2 + \tau\sum_{j=1}^N \|\widetilde{u}^j - \widehat{u}^j\|_{L^2(\Omega)}^2 &\le C(h^{2s} + \Omega_D(h)), \\ \max_{1\le m\le M}\Big\|\tau\sum_{j=1}^m \nabla(\widetilde{u}^j - \widehat{u}^j)\Big\|_{L^2(\Omega)}^2 &\le Ch^{-2}(h^{2s} + \Omega_D(h)), \\ \tau\sum_{j=1}^N \|\widetilde{u}^j - \widehat{u}^j\|_{L^2(\Gamma)}^2 &\le Ch^{-1}(h^{2s} + \Omega_D(h)). \end{aligned}$$

Next, recalling $(\widetilde{I}_\varepsilon)^\tau$ from §3 and combining it with $(I_\varepsilon)_h^\tau$ and $(\widetilde{P})_h$, we see that the function $U_h^j - \widehat{u}^j \in V_h^0$ satisfies the relation

$$\begin{aligned} &\Big(\frac{1}{\tau}(\theta(U_h^j) - \theta(u^j)) - \frac{1}{\tau}(\theta(U_h^{j-1}) - \theta(u^{j-1})), \varphi\Big) \\ &\qquad +\Big\langle\frac{1}{\tau}(\beta(U_h^j) - \beta(u^j)) - \frac{1}{\tau}(\beta(U_h^{j-1}) - \beta(u^{j-1})), \varphi\Big\rangle + (\nabla(U_h^j - \widehat{u}^j), \nabla\varphi) \\ &= (f(\theta(U_h^j)) - \widetilde{f}^j, \varphi) - (b(\theta(U_h^j)) - \widetilde{b}^j, \nabla\varphi) + \langle g(\beta(U_h^j)) - \widetilde{g}^j, \varphi\rangle, \end{aligned}$$

where u^j, $\widetilde{f}^j$, $\widetilde{b}^j$, $\widetilde{g}^j$ appear as in §3.

We now choose $\varphi = \varphi^j = \sum_{i=j}^m (U_h^i - \widehat{u}^i)$ and sum with respect to j from 1 to m ($1 \le m \le N$). We then have

$$\begin{aligned}
&\sum_{j=1}^m \Big(\frac{1}{\tau}(\theta(U_h^j) - \theta(u^j)) - \frac{1}{\tau}(\theta(U_h^{j-1}) - \theta(u^{j-1})), \sum_{i=j}^m (U_h^i - \widehat{u}^i)\Big) \\
&= -\frac{1}{\tau}(\theta(U_h^0) - \theta(u^0), \sum_{i=1}^m (U_h^i - \widehat{u}^i)) + \frac{1}{\tau}\sum_{j=1}^m (\theta(U_h^j) - \theta(u^j), U_h^j - u^j + u^j - \widehat{u}^j) \\
&\ge (\mu - \lambda)\tau^{-1}\sum_{j=1}^m \|\theta(U_h^j) - \theta(u^j)\|^2_{L^2(\Omega)} - \lambda\|\varphi^1\|^2_{L^2(\Omega)} \\
&\quad - C_\lambda[\tau^{-2}(\Omega_0(h) + h^{2s} + \Omega_D(h)) + \varepsilon^{-1}],
\end{aligned}$$

since by Lemma 4.1 and Corollary 2.1

$$\begin{aligned}
\tau\sum_{j=1}^m \|u^j - \widehat{u}^j\|^2_{L^2(\Omega)} &\le 2\Big(\tau\sum_{j=1}^m \|u^j - \widetilde{u}^j\|^2_{L^2(\Omega)} + \tau\sum_{j=1}^m \|\widetilde{u}^j - \widehat{u}^j\|^2_{L^2(\Omega)}\Big) \\
&\le C(\varepsilon^{-1}\tau^2 + h^{2s} + \Omega_D(h)).
\end{aligned}$$

Similarly,

$$\begin{aligned}
&\sum_{j=1}^m \Big\langle\frac{1}{\tau}(\beta(U_h^j) - \beta(u^j)) - \frac{1}{\tau}(\beta(U_h^{j-1}) - \beta(u^{j-1})), \sum_{i=j}^m (U_h^i - \widehat{u}^i)\Big\rangle \\
&\quad \ge (\mu - \lambda)\tau^{-1}\sum_{j=1}^m \|\beta(U_h^j) - \beta(u^j)\|^2_{L^2(\Gamma)} - \lambda\|\varphi^1\|^2_{L^2(\Gamma)} \\
&\quad - C_\lambda[\tau^{-2}(\Omega_0(h) + h^{2s-1} + h^{-1}\Omega_D(h)) + \varepsilon^{-1}].
\end{aligned}$$

Moreover,

$$\begin{aligned}
&\sum_{j=1}^m (\nabla(U_h^j - \widehat{u}^j), \sum_{i=j}^m \nabla(U_h^i - \widehat{u}^i)) \ge \frac{1}{2}\Big\|\sum_{j=1}^m \nabla(U_h^j - \widehat{u}^j)\Big\|^2_{L^2(\Omega)} = \frac{1}{2}\|\nabla\varphi^1\|^2_{L^2(\Omega)}, \\
&\sum_{j=1}^m (f(\theta(U_h^j)) - \widetilde{f}^j, \varphi^j) \le \lambda\tau^{-1}\sum_{j=1}^m \|\theta(U_h^j) - \theta(u^j)\|^2_{L^2(\Omega)} + C_\lambda\Big(1 + \sum_{j=1}^m \|\varphi^j\|^2_{L^2(\Omega)}\Big), \\
-&\sum_{j=1}^m (b(\theta(U_h^j)) - \widetilde{b}^j, \nabla\varphi^j) \le \lambda\tau^{-1}\sum_{j=1}^m \|\theta(U_h^j) - \theta(u^j)\|^2_{L^2(\Omega)} + C_\lambda\Big(1 + \sum_{j=1}^m \|\nabla\varphi^j\|^2_{L^2(\Omega)}\Big), \\
&\sum_{j=1}^m \langle g(\beta(U_h^j)) - \widetilde{g}^j, \varphi^j\rangle \le \lambda\tau^{-1}\sum_{j=1}^m \|\beta(U_h^j) - \beta(u^j)\|^2_{L^2(\Gamma)} + C_\lambda\Big(1 + \sum_{j=1}^m \|\varphi^j\|^2_{L^2(\Gamma)}\Big).
\end{aligned}$$

Thus,

$$\begin{aligned}
&\tau^{-1}\sum_{j=1}^m \|\theta(U_h^j) - \theta(u^j)\|^2_{L^2(\Omega)} + \tau^{-1}\sum_{j=1}^m \|\beta(U_h^j) - \beta(u^j)\|^2_{L^2(\Gamma)} + \|\nabla\varphi^1\|^2_{L^2(\Omega)} \\
&\le C\Big[\varepsilon^{-1} + \tau^{-2}h^{-1}(h^{2s} + \Omega_D(h)) + \tau^{-2}\Omega_0(h) + \sum_{j=1}^m \|\nabla\varphi^j\|^2_{L^2(\Omega)}\Big],
\end{aligned}$$

where $\varphi^j = \sum_{i=j}^m (U_h^i - \widehat{u}^i)$. Finally, by a Gronwall technique we arrive at the following estimates.

Theorem 4.1

$$\tau \sum_{j=1}^{N} \|\theta_\varepsilon(U_h^j) - \theta_\varepsilon(u_\varepsilon^j)\|_{L^2(\Omega)}^2 + \tau \sum_{j=1}^{N} \|\beta_\varepsilon(U_h^j) - \beta_\varepsilon(u_\varepsilon^j)\|_{L^2(\Gamma)}^2$$
$$\leq C\{\varepsilon^{-1}\tau^2 + h^{-1}(h^{2s} + \Omega_D(h)) + \Omega_0(h)\},$$
$$\max_{1\leq m\leq N} \left\| \nabla\left(\tau \sum_{j=1}^{m} U_h^j - \int_0^{m\tau} u_\varepsilon\right)\right\|_{L^2(\Omega)}^2 \leq C\{\varepsilon^{-1}\tau^2 + h^{-2}(h^{2s} + \Omega_D(h)) + \Omega_0(h)\}.$$

Define the function $u_{\varepsilon h}^\tau : [0,T] \to H^1(\Omega)$ as follows:

$$u_{\varepsilon h}^\tau(t) = U_h^j \quad \text{if } t \in ((j-1)\tau, j\tau].$$

Then, Theorem 4.1 presents an estimate of the error between $u_{\varepsilon h}^\tau$ and u_ε, while Theorem 2.1 gives an error estimate between u_ε and u, the weak solution of problem (I). Consequently, we have proved the following theorem.

Theorem 4.2

$$\int_0^T \|\theta_\varepsilon(u_{\varepsilon h}^\tau) - \theta(u)\|_{L^2(\Omega)}^2 + \int_0^T \|\beta_\varepsilon(u_{\varepsilon h}^\tau) - \beta(u)\|_{L^2(\Gamma)}^2$$
$$\leq C\{\varepsilon + \varepsilon^{-1}\tau^2 + h^{-1}(h^{2s} + \Omega_D(h)) + \Omega_0(h)\},$$
$$\sup_{0\leq t\leq T} \left\| \nabla \int_0^t (u_{\varepsilon h}^\tau - u)\right\|_{L^2(\Omega)}^2 \leq C\{\varepsilon + \varepsilon^{-1}\tau^2 + h^{-2}(h^{2s} + \Omega_D(h)) + \Omega_0(h)\}.$$

It follows from the hypotheses $(H_1)'$ and (H_3) and the approximation property of the finite element space V_h (see Ciarlet [2]) that we can choose u_{Dh} and u_{0h} such that

$$h^{-1}\Omega_D(h) \sim \Omega_0(h) \sim h^{2r-1}.$$

Then one estimate of Theorem 4.2 reduces to

$$\int_0^T \|\theta_\varepsilon(u_{\varepsilon h}^\tau) - \theta(u)\|_{L^2(\Omega)}^2 + \int_0^T \|\beta_\varepsilon(u_{\varepsilon h}^\tau) - \beta(u)\|_{L^2(\Gamma)}^2 \leq C(\varepsilon + \varepsilon^{-1}\tau^2 + h^{2s-1}),$$

since $s \leq r$. If $\beta(z) \equiv 0$, then h^{2s-1} can be replaced by h^{2s}.

Acknowledgement. This work was supported by the National Foundation of Natural Science, P. R. China, and completed in 1991-92 when the author was visiting the Faculty of Mathematics and Mechanics, Moscow University, Moscow, Russia.

REFERENCES

[1] Alt H. W., Luckhaus S. Quasilinear elliptic-parabolic differential equations. *Math. Z.*, 183:311–341, 1983.

[2] Ciarlet P. G. *The Finite Element Method for Elliptic Problems.* North Holland, Amsterdam, 1978.

[3] Czou Y. L. Boundary value problems for nonlinear parabolic equations. *Mat. Sb.*, 47:431–484, 1959.

[4] Li J. G., Su N. The solution of an infiltration problem with ponded surface flux condition. *Acta Math. Appl. Sinica*, 2:54–65, 1985. English series.

[5] Nochetto R. H. Error estimates for two phase Stefan problems in several space variables, I: linear boundary conditions. *Calcolo*, 22:457–499, 1985.

[6] Verdi C. Optimal error estimates for an approximation of degenerate parabolic problems. *Numer. Funct. Anal. Optimization*, 9:657–670, 1987.

Titles previously published in the series

INTERNATIONAL SERIES OF NUMERICAL MATHEMATICS
BIRKHÄUSER VERLAG

ISNM 89 **K. Gürlebeck / W. Sprössig:** Quaternionic Analysis and Elliptic Boundary Value Problems. 1990 (ISBN 3-7643-2382-5)

ISNM 90 **C.K. Chui / W. Schempp / K. Zeller (Eds):** Multivariate Approximation Theory IV. Proceedings of the Conference at the Mathematical Research Institute at Oberwolfach, Black Forest, February 12–18, 1989. 1989 (ISBN 3-7643-2384-1)

ISNM 91 **Kappel, F. / Schappacher, W. / K. Kunisch (Eds):** Control and Estimation of Distributed Parameter Systems. 4th International Conference on Control of Distributed Parameter Systems, Vorau, July 10–16, 1988. 1989 (ISBN 3-7643-2345-0)

ISNM 92 **H.D. Mittelmann / D. Roose (Eds):** Continuation Techniques and Bifurcation Problems. 1990 (ISBN 3-7643-2397-3)

ISNM 93 **R.E. Bank, R. Bulirsch, K. Merten (Eds):** Mathematical Modelling and Simulation of Electric Circuits and Semiconductor Devices. 1990 (ISBN 3-7643-2439-2)

ISNM 94 **W. Haussmann, K. Jetter (Eds):** Multivariate Approximation and Interpolation. 1990 (ISBN 3-7643-2450-3)

ISNM 95 **K.-H. Hoffmann, J. Sprekels (Eds):** Free Boundary Value Problems. 1990 (ISBN 3-7643-2474-0)

ISNM 96 **J. Albrecht, L. Collatz, P. Hagedorn, W. Velte (Eds):** Numerical Treatment of Eigenvalue Problems, Vol. 5. 1991 (ISBN 3-7643-2575-5)

ISNM 97 **R.U. Seydel, F.W. Schneider, T.G. Küpper, H. Troger (Eds):** Bifurcation and Chaos: Analysis, Algorithms, Applications. 1991 (ISBN 3-7643-2593-3)

ISNM 98 **W. Hackbusch, U. Trottenberg (Eds):** Multigrid Methods III. 1991 (ISBN 3-7643-2632-8)

ISNM 99 **P. Neittaanmäki (Ed.):** Numerical Methods for Free Boundary Problems. 1991 (ISBN 3-7643-2641-7)

ISNM 100 **W. Desch, F. Kappel, K. Kunisch (Eds):** Estimation and Control of Distributed Parameter Systems. 1991 (ISBN 3-7643-2676-X)

ISNM 101 **G. Del Piero, F. Maceri (Eds):** Unilateral Problems in Structural Analysis IV. 1991 (ISBN 3-7643-2487-2)

ISNM 102 **U. Hornung, P. Kotelenez, G. Papanicolaou (Eds):** Random Partial Differential Equations. 1991 (ISBN 3-7643-2688-3)

ISNM 103 **W. Walter (Ed.):** General Inequalities 6. 1992 (ISBN 3-7643-2737-5)

ISNM 104 **E. Allgower, K. Böhmer, M. Golubitsky (Eds):** Bifurcation and Symmetry. 1992 (ISBN 3-7643-2739-1)

ISNM 105 **D. Braess, L.L. Schumaker (Eds):** Numerical Methods in Approximation Theory, Vol. 9. 1992 (ISBN 3-7643-2746-4)

ISNM 106 **S.N. Antontsev, K.-H. Hoffmann, A.M. Khludnev (Eds):** Free Boundary Problems in Continuum Mechanics. 1992 (ISBN 3-7643-2784-7)

ISNM 107 **V. Barbu, F.J. Bonnans, D. Tiba (Eds):** Optimization, Optimal Control and Partial Differential Equations. 1992 (ISBN 3-7643-2788-X)

ISNM 108 **H. Antes, P.D. Panagiotopoulos:** The Boundary Integral Approach to Static and Dynamic Contact Problems. Equality and Inequality Methods. 1992 (ISBN 3-7643-2592-5)

ISNM 109 **A.G. Kuz'min:** Non-Classical Equations of Mixed Type and their Applications in Gas Dynamics. 1992 (ISBN 3-7643-2573-9)

ISNM 110 **H.R.E.M. Hörnlein, K. Schittkowski (Eds):** Software Systems for Structural Optimization. 1992 (ISBN 3-7643-2836-3)

ISNM 111 **R. Burlisch, A. Miele, J. Stoer, K.H. Well:** Optimal Control. 1993 (ISBN 3-7643-2887-8)

ISNM 112 **H. Braess, G. Hämmerlin (Eds):** Numerical Integration IV. Proceedings of the Conference at the Mathematical Research Institute at Oberwolfach, November 8–14, 1992. 1993 (ISBN 3-7643-2922-X)

ISNM 113 **L. Quartapelle:** Numerical Solution of the Incompressible Navier-Stokes Equations. 1993 (ISBN 3-7643-2935-1)

ISNM 114 **J. Douglas, U. Hornung (Eds):** Flow in Porous Media. 1993 (ISBN 3-7643-2949-1)

Zeitfracht Medien GmbH
Ferdinand-Jühlke-Straße 7
99095 Erfurt, Deutschland
produktsicherheit@kolibri360.de